STUDENT MATH JOURNAL
ANSWER BOOK
VOLUME 1

Everyday
Mathematics®

The University of Chicago School Mathematics Project

Mc
Graw
Hill
Education

The University of Chicago School Mathematics Project

Max Bell, Director, *Everyday Mathematics* First Edition; James McBride, Director, *Everyday Mathematics* Second Edition; Andy Isaacs, Director, *Everyday Mathematics* Third, CCSS, and Fourth Editions; Amy Dillard, Associate Director, *Everyday Mathematics* Third Edition; Rachel Malpass McCall, Associate Director, *Everyday Mathematics* CCSS and Fourth Editions; Mary Ellen Dairyko, Associate Director, *Everyday Mathematics* Fourth Edition

Authors
Robert Balfanz*, Max Bell, John Bretzlauf, Sarah R. Burns**, William Carroll*, Amy Dillard, Robert Hartfield, Andy Isaacs, James McBride, Kathleen Pitvorec, Denise A. Porter‡, Peter Saecker, Noreen Winningham†

*First Edition only
** Fourth Edition only
†Third Edition only
‡Common Core State Standards Edition only

Fourth Edition Grade 5

Team Leader
Sarah R. Burns

Writers
Melanie S. Arazy, Rosalie A. DeFino, Allison M. Greer, Kathryn M. Rich, Linda M. Sims

Open Response Team
Catherine R. Kelso, Leader; Emily Korzynski

Differentiation Team
Ava Belisle-Chatterjee, Leader; Martin Gartzman, Barbara Molina, Anne Sommers

Digital Development Team
Carla Agard-Strickland, Leader; John Benson, Gregory Berns-Leone, Juan Camilo Acevedo

Virtual Learning Community
Meg Schleppenbach Bates, Cheryl G. Moran, Margaret Sharkey

Technical Art
Diana Barrie, Senior Artist; Cherry Inthalangsy

UCSMP Editorial
Don Reneau, Senior Editor; Rachel Jacobs, Elizabeth Olin, Kristen Pasmore, Loren Santow

Field Test Coordination
Denise A. Porter, Angela Schieffer, Amanda Zimolzak

Field Test Teachers
Diane Bloom, Margaret Condit, Barbara Egofske, Howard Gartzman, Douglas D. Hassett, Aubrey Ignace, Amy Jarrett-Clancy, Heather L. Johnson, Jennifer Kahlenberg, Deborah Laskey, Jennie Magiera, Sara Matson, Stephanie Milzenmacher, Sunmin Park, Justin F. Rees, Toi Smith

Contributors
John Benson, Jeanne Di Domenico, James Flanders, Fran Goldenberg, Lila K. S. Goldstein, Deborah Arron Leslie, Sheila Sconiers, Sandra Vitantonio, Penny Williams

Center for Elementary Mathematics and Science Education Administration
Martin Gartzman, Executive Director; Meri B. Fohran, Jose J. Fragoso, Jr., Regina Littleton, Laurie K. Thrasher

External Reviewers
The *Everyday Mathematics* authors gratefully acknowledge the work of the many scholars and teachers who reviewed plans for this edition. All decisions regarding the content and pedagogy of *Everyday Mathematics* were made by the authors and do not necessarily reflect the views of those listed below.

Elizabeth Babcock, California Academy of Sciences; Arthur J. Baroody, University of Illinois at Urbana-Champaign and University of Denver; Dawn Berk, University of Delaware; Diane J. Briars, Pittsburgh, Pennsylvania; Kathryn B. Chval, University of Missouri–Columbia; Kathleen Cramer, University of Minnesota; Ethan Danahy, Tufts University; Tom de Boor, Grunwald Associates; Louis V. DiBello, University of Illinois at Chicago; Corey Drake, Michigan State University; David Foster, Silicon Valley Mathematics Initiative; Funda Gönülateş, Michigan State University; M. Kathleen Heid, Pennsylvania State University; Natalie Jakucyn, Glenbrook South High School, Glenview, IL; Richard G. Kron, University of Chicago; Richard Lehrer, Vanderbilt University; Susan C. Levine, University of Chicago; Lorraine M. Males, University of Nebraska-Lincoln; Dr. George Mehler, Temple University and Central Bucks School District, Pennsylvania; Kenny Huy Nguyen, North Carolina State University; Mark Oreglia, University of Chicago; Sandra Overcash, Virginia Beach City Public Schools, Virginia; Raedy M. Ping, University of Chicago; Kevin L. Polk, Aveniros LLC; Sarah R. Powell, University of Texas at Austin; Janine T. Remillard, University of Pennsylvania; John P. Smith III, Michigan State University; Mary Kay Stein, University of Pittsburgh; Dale Truding, Arlington Heights District 25, Arlington Heights, Illinois; Judith S. Zawojewski, Illinois Institute of Technology

Note
Many people have contributed to the creation of *Everyday Mathematics*. Visit http://everydaymath.uchicago.edu/authors/ for biographical sketches of *Everyday Mathematics* 4 staff and copyright pages from earlier editions.

www.everydaymath.com

Send all inquiries to:
McGraw-Hill Education
8787 Orion Place
Columbus, OH 43240

ISBN: 978-0-02-137609-4
MHID: 0-02-137609-3

Printed in the United States of America.

1 2 3 4 5 6 7 8 9 QVS 20 19 18 17 16 15

Contents

Unit 1

Welcome to *Fifth Grade Everyday Mathematics* 1

Student Reference Book Scavenger Hunt 2

Math Boxes 1-1 4

Areas of Rectangles 5

Finding Areas of Rectangles 6

Math Boxes 1-2 7

Area in Two Units 8

Math Boxes 1-3 9

A Tiling Strategy 10

Solving Area Problems 11

Math Boxes 1-4 12

Comparing Volume 13

Math Boxes 1-5 14

Packing Prisms to Measure Volume 15

More Areas of Rectangles 16

Math Boxes 1-6 17

Measuring Volume with Cubes 18

Math Boxes 1-7 20

Converting Measurements 21

Cube-Stacking Problems Using Layers 22

Math Boxes 1-8: Preview for Unit 2 24

Finding Volume Using Formulas 25

Writing and Interpreting Expressions 26

Math Boxes 1-9 27

Converting Cubic Units 28

More Cube-Stacking Problems 30

Math Boxes 1-10 31

Estimating Volumes of Instrument Cases 32

Math Boxes 1-11 33

Understanding Grouping Symbols 34

Math Boxes 1-12 35

Math Boxes 1-13: Preview for Unit 2 36

Unit 2

Place-Value Relationships 37

Finding Volumes of Rectangular Prisms 39

Math Boxes 2-1 40

Powers of 10 and Exponential Notation 41

Expanded Form with Powers of 10 42

Solving a Real-World Volume Problem 43

Math Boxes 2-2 44

Estimating with Powers of 10 45

Writing and Comparing Expressions 46

Math Boxes 2-3 47

Practicing Multiplication Strategies 48

Math Boxes 2-4 49

Practicing with Powers of 10 50

Math Boxes 2-5 51

Unit Conversion Number Stories 52

Math Boxes 2-6 53

Estimating and Multiplying with 2-Digit Numbers 54

Math Boxes 2-7 55

Choosing Multiplication Strategies 56

Math Boxes 2-8 57

Invitations 58

Math Boxes 2-9 59

Using Multiples to Rename Dividends 60

Practicing Unit Conversions 61

Math Boxes 2-10: Preview for Unit 3 62

Partial-Quotients Division 63

Math Boxes 2-11 64

Partial Quotients with Multiples 65

Math Boxes 2-12 66

Math Boxes 2-13 67

Interpreting Remainders 68

Math Boxes 2-14: Preview for Unit 3 70

Unit 3

Solving Fair Share Number Stories 71

Multiplication and Division 72

Math Boxes 3-1 73

Writing Division Number Stories 74

More Practice with Fair Shares 75

Math Boxes 3-2 76

Division Number Stories with Remainders 77

Math Boxes 3-3 79

Fractions on a Number Line 80

Math Boxes 3-4 82

Division Top-It 83

Math Boxes 3-5 84

Checking for Reasonable Answers 85

Interpreting Remainders in Number Stories 86

Math Boxes 3-6 87

Using Benchmarks to Make Estimates 88

Math Boxes 3-7 89

Renaming Fractions and Mixed Numbers 90

Connecting Fractions and Division 91

Math Boxes 3-8 92

Addition and Subtraction Number Stories 93

Math Boxes 3-9 94

Adding Fractions with Circle Pieces 95

Explaining Place-Value Patterns 96

Math Boxes 3-10: Preview for Unit 4 98

Renaming Fractions and Mixed Numbers 99

Math Boxes 3-11 100

Solving Fraction Number Stories 101

Practicing Division 103

Math Boxes 3-12 104

Fraction-Of Problems 105

Using Models to Estimate Volumes 106

Math Boxes 3-13 107

More Fraction-Of Problems 108

Math Boxes 3-14 109

Math Boxes 3-15: Preview for Unit 4 110

Unit 4

Math Boxes 4-1 111

Reading and Writing Decimals 112

Representing Decimals 114

Fraction Number Stories 116

Math Boxes 4-2 117
Writing Decimals in Expanded Form 118
Representing Decimals in Expanded Form 119
Math Boxes 4-3 120
Interpreting Remainders 121
Math Boxes 4-4 122
Identifying the Closer Number 123
Rounding Decimals 124
Rounding Decimals in Real-World Contexts 125
Math Boxes 4-5 126
Locating Cities on a Map of Ireland 127
Plotting Points on a Coordinate Grid 128
How Much Soil? 129
Math Boxes 4-6 130
Town Map 131
Practice with U.S. Traditional Multiplication 132
Math Boxes 4-7 133
Graphing Sailboats 134
A New Sailboat Rule 136
Math Boxes 4-8 137
Graphing Data as Ordered Pairs 138
Forming and Graphing Ordered Pairs 139
Math Boxes 4-9 140
Logo 141
Math Boxes 4-10: Preview for Unit 5 143
Decimal Addition and Subtraction with Grids 144
Math Boxes 4-11 145
Using Algorithms to Add Decimals 146
Math Boxes 4-12 147
Using Algorithms to Subtract Decimals 148
Math Boxes 4-13 149
Finding Areas of New Floors 150
Math Boxes 4-14 151
Math Boxes 4-15: Preview for Unit 5 152

Activity Sheets

Rectangular Prism Patterns . . . . Activity Sheet 1
More Rectangular Prism Patterns . . . . Activity Sheet 2
Prism Pile-Up Cards . . . . Activity Sheet 3

Prism Pile-Up Cards (continued) . . . . Activity Sheet 4

Fraction Circle Pieces 1 . . . . Activity Sheet 5

Fraction Circle Pieces 2 . . . . Activity Sheet 6

Fraction Circle Pieces 3 . . . . Activity Sheet 7

Fraction Cards 1 . . . . Activity Sheet 8

Fraction Cards 2 . . . . Activity Sheet 9

Fraction Cards 3 . . . . Activity Sheet 10

Fraction Cards 4 . . . . Activity Sheet 11

Fraction Cards 5 . . . . Activity Sheet 12

Fraction Cards 6 . . . . Activity Sheet 13

Fraction Of Fraction Cards (Set 1) . . . . Activity Sheet 14

Fraction Of Whole Cards . . . . Activity Sheet 15

Blank Cards . . . . Activity Sheet 16

*Activity Sheets do not appear in the *Math Journal Answer Book*.

Welcome to *Fifth Grade Everyday Mathematics*

This year in math class you will continue to build on the mathematical skills and ideas you have learned in previous years. You will learn new mathematics and think about the importance of mathematics in your life now and how math will be useful to you in the future. Many of the new ideas you learn this year will be ones that your parents, or even your older brothers and sisters, may not have learned until much later than fifth grade. The authors of *Everyday Mathematics* believe that today's fifth graders are able to learn more and do more than fifth graders in the past. They think that mathematics is fun and they think you will find it enjoyable too.

Here are some of the things you will do in *Fifth Grade Everyday Mathematics*:

- Extend your understanding of place value to decimals and use what you learn to explain how our place-value system works.

- Review and extend your skills doing arithmetic, using a calculator, and thinking about problems and their solutions. You will add, subtract, multiply, and divide whole numbers and decimals.

- Use your knowledge of fractions and operations to compute with fractions. You will think about how adding, subtracting, multiplying, and dividing fractions is similar to and different from doing the same computations with whole numbers and decimals.

- Explore the concept of volume. You will learn how volume differs from other measurements you have studied. You will find the volume of 3-dimensional figures in multiple ways, and you will develop strategies for finding the volume of rectangular prisms. Look at journal page 2. Without telling anyone, write the number one hundred twelve in the top right-hand corner of the page.

- Learn about coordinate grids and find out how graphing can help you solve mathematical and real-world problems.

- Deepen your understanding of 2-dimensional figures, their attributes, and how different 2-dimensional figures are related to each other.

We want you to become better at using mathematics so you can better understand your world. We hope you enjoy the activities in *Fifth Grade Everyday Mathematics* and that they help you appreciate the beauty and usefulness of mathematics in your daily life.

Solve the problems on this page and page 3. Use your *Student Reference Book* to find information about each problem. Record the page numbers.

112

| | Problem Points | Page Points |

1 5 meters = __500__ centimeters   _____ _____

 SRB
page __213, 328__

2 300 mm = __30__ cm   _____ _____

 SRB
page __213, 328__

3 Solve.   _____ _____

(15 − 4) * 3 = __33__

25 + (47 − 18) = __54__

 SRB
page __42__

4 Write the value of the 5 in each of the following numbers.   _____ _____

9,652 __50__

15,690 __5,000__

1,052,903 __50,000__

 SRB
page __66–67__

5 Name two fractions equivalent to $\frac{4}{6}$. Sample answers: _____ _____

__$\frac{2}{3}$__ and __$\frac{12}{18}$__

 SRB
page __166, 168–170__

6 460 ÷ 5 = __92__   _____ _____

 SRB
page __108__

| | Problem Points | Page Points |

7 **a.** What is the definition of a trapezoid? _____ _____

A quadrilateral that has at least one pair of parallel sides.

b. Draw two different trapezoids. Sample answers:

 page 268

8 What materials do you need to play *Name That Number*? _____ _____

1 complete deck of number cards

 page 315

Record your scavenger hunt scores in the table below. Then calculate the totals.

Problem Number	Problem Points	Page Points	Total Points
1			
2			
3			
4			
5			
6			
7			
8			
Total Points			

5.G.3, SMP5

3

Math Boxes

Math Boxes

1 Next to each *Student Reference Book* icon in Problems 2–5, write the SRB page numbers where you can find information about each problem.

2 Solve.

a. $(25 - 5) * 4 = $ __80__

b. $25 - (5 * 4) = $ __5__

__42__

3 Complete.

a. 1 foot = __12__ inches

b. A person who is 6 feet tall is

__72__ inches tall.

c. 1 yard = __3__ feet

d. A person who ran 300 yards

ran __900__ feet.   215–216, 328

4 Without calculating, circle ALL of the expressions that are greater than 2 + 8.

(A.) $4 + (2 + 8)$

B. $2 + 8 - 5$

(C.) $2 + 8 + 10$

D. $8 + 2$

__46__

5 **Writing/Reasoning** Explain how you solved Problem 2b.

Sample answer: I knew I had to start with the operations inside the parentheses, so I did 5 × 4 first and got 20. Then I subtracted 20 from 25 and got 5.

__42__

Areas of Rectangles

Write two important facts that you learned about *area* after reading
Student Reference Book, page 221.

SRB
221

① Answers vary.

_____

_____

② _____

_____

In Problems 3 and 4 each grid square is 1 square unit.
Find the area of each rectangle. Don't forget to include a unit.

③

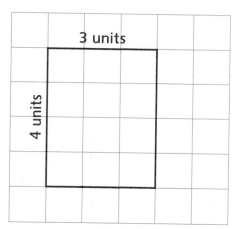

3 units

4 units

④

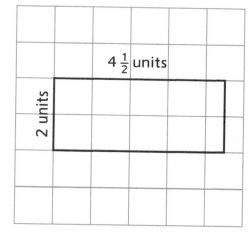

$4\frac{1}{2}$ units

2 units

Area: __12 square units__

Area: __9 square units__

⑤ Think about how you found the area of the rectangle in Problem 3 and how you found the
area of the rectangle in Problem 4. What was the same? What was different? Record your
thoughts. Be prepared to share them with the class.

Sample answer: In both problems, I thought about counting
squares. In Problem 3 I just had to count whole squares.
In Problem 4 there were some squares that weren't whole,
so I had to think about how to add those into the area.

5.NF.4, 5.NF.4b, SMP1

Find the area of each rectangle.
Write a number sentence to show your thinking.

Number sentences are sample answers.

1

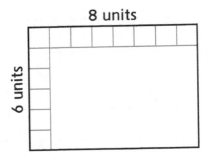

8 units

6 units

Area = __48__ square units

$6 * 8 = 48$

(number sentence)

2

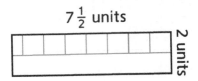

$7\frac{1}{2}$ units

2 units

Area = __15__ square units

$7\frac{1}{2} + 7\frac{1}{2} = 15$

(number sentence)

3

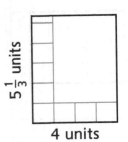

$5\frac{1}{3}$ units

4 units

Area = __$21\frac{1}{3}$__ square units

$5\frac{1}{3} + 5\frac{1}{3} + 5\frac{1}{3} + 5\frac{1}{3} = 21\frac{1}{3}$

(number sentence)

Try This

4

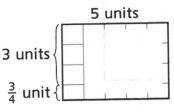

5 units

3 units

$\frac{3}{4}$ unit

Area = __$18\frac{3}{4}$__ square units

$5 * 3\frac{3}{4} = 18\frac{3}{4}$

(number sentence)

5 Explain the strategy you used to find the area of the rectangle in Problem 3. Use words like *row*, *column*, *square unit*, and *partial square* to help make your thinking clear.

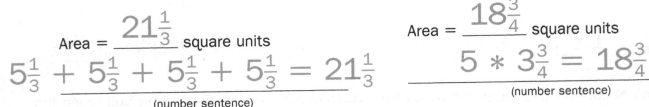

Sample answer: I thought about making copies of the column to fill up the rectangle. Each column had $5\frac{1}{3}$ square units, and I thought about 4 copies. $4 * 5\frac{1}{3}$ is the same as $5\frac{1}{3} + 5\frac{1}{3} + 5\frac{1}{3} + 5\frac{1}{3} = 20\frac{4}{3}$, or $21\frac{1}{3}$.

Math Boxes

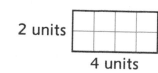

---

1 Place parentheses in a different place in each problem. Then solve.

$(6 * 3) + 8 =$ ___26___

$6 * (3 + 8) =$ ___66___

SRB
42

---

2 Find the area of the rectangle.

2 units

4 units

Area = ___8___ square units

SRB
224-225

---

3 Write an expression for the number story.

Alexa had 3 boxes of 10 shells. She bought shells at a store to double her collection.

Sample answer:
$3 * 10 * 2$

SRB
38, 44

---

4 Complete the table.

Feet	Inches
1	12
10	120
5	60
$3\frac{1}{2}$	42

SRB
215-216, 328

---

5 Jamir has 137 dollars in the bank and 25 dollars at home. Ricky has twice as much money as Jamir.

Circle the expression that represents Ricky's money.

a. $2 * 137 + 25$

b. $2 * (137 + 25)$

SRB
42, 46

---

6 Give the value of the 4 in each number.

a. 42,671 ___40,000___

b. 64,671 ___4,000___

c. 62,471 ___400___

d. 62,641 ___40___

e. 62,674 ___4___

SRB
66-67

---

① 5.OA.1 ② 5.NF.4, 5.NF.4b ③ 5.OA.2 ④ 5.MD.1
⑤ 5.OA.1, 5.OA.2 ⑥ 5.NBT.1

7

Area in Two Units

Noah has a piece of paper with an area of 1 square foot. He plans to paint a design on it using smaller squares, each with an area of 1 square inch. Below is the plan for his design.

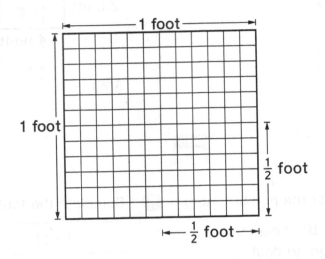

1. Use the picture to find the number of smaller squares (with an area of 1 square inch) that Noah will use in his design.

    He will use ___144___ square inches in his design.

2. Noah wants to make a second, smaller design in a square with a side length of $\frac{1}{2}$ foot. Use the picture to find the number of squares (with an area of 1 square inch) that Noah will use in the second design.

    He will use ___36___ square inches in his second design.

Math Boxes

1 Where in the *Student Reference Book* would you look to find the definition of *area*?

Circle the best answer.

a. Table of contents

b. Index

c. Glossary *(circled)*

d. Games section

e. All of the above

2 Solve.

a. $(4 * 12) + 8 =$ __56__

b. __4__ $= 32 / (16 \div 2)$

c. __8__ $= (32 \div 8) * 2$

SRB
42

3 Draw lines to match each measurement with its equivalent.

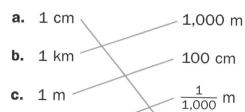

a. 1 cm   1,000 m

b. 1 km   100 cm

c. 1 m $\frac{1}{1,000}$ m

d. 1 mm $\frac{1}{100}$ m

SRB
213, 328

4 Two friends were playing a game and recorded their scores below.

Player 1: $42 + 51$ points

Player 2: $4 + (42 + 51)$ points

Who has more points? __Player 2__

SRB
46

5 **Writing and Reasoning** Did you have to calculate the scores to find out who had more points in Problem 4? Why or why not?

Sample answer: No. I know Player 2 has the higher score because her score is 4 more than Player 1's score.

SRB
46

① SMP5 ② 5.OA.1 ③ 5.MD.1 ④ 5.OA.2
⑤ 5.OA.2, SMP2, SMP3

A Tiling Strategy

SRB
226

Math Message

1 In Lesson 1-3 you found that 4 squares with a side length of $\frac{1}{2}$ foot fit into 1 square foot. How many squares with a side length of $\frac{1}{3}$ foot do you think would fit into 1 square foot? Use the pictures below to help. Be ready to explain how you found your answer.

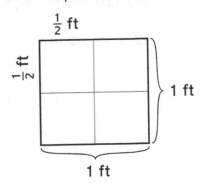

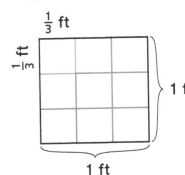

___9___ squares with a side length of $\frac{1}{3}$ foot fit into one square foot.

---

2 Roger's shower is $2\frac{2}{3}$ feet wide and 3 feet long. He is going to cover the floor of the shower with square tiles that measure $\frac{1}{3}$ foot on each side.

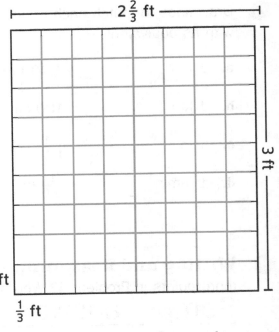

a. How many tiles will Roger need to cover the shower floor? Use the picture to help.

___72___ tiles

b. How many of Roger's tiles does it take to cover 1 square foot?

___9___ tiles

c. Use your answers to Parts a and b to find the area of Roger's shower floor in square feet.

___8___ square feet

_____ $72 / 9 = 8$ _____
(number sentence)

3 Summarize the strategy you used to find the area of Roger's shower floor. Sample answer: First I found that 72 tiles covered the floor. Then I figured out that 9 tiles cover 1 square foot. I divided 72 by 9 to get 8 square feet.

Solving Area Problems

1   Anna is covering the top of her jewelry box with glass tiles that are $\frac{1}{2}$ inch long and $\frac{1}{2}$ inch wide. The top of the jewelry box is $3\frac{1}{2}$ inches by 2 inches.

SRB
226

a. How many tiles will she need to cover the top of the box? Use the picture to help.

____28____ tiles

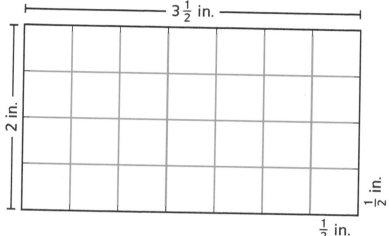

$3\frac{1}{2}$ in.

2 in.

$\frac{1}{2}$ in.

$\frac{1}{2}$ in.

b. How many of Anna's tiles does it take to cover 1 square inch?

____4____ tiles

c. Use your answers to Parts a and b to find the area of the top of the jewelry box in square inches. ____7____ square inches

____28 / 4 = 7____
(number sentence)

2   Deshawn is covering a 4-yard by $1\frac{3}{4}$-yard section of his bedroom wall with decorative tiles. The tiles are $\frac{1}{4}$ yard by $\frac{1}{4}$ yard.

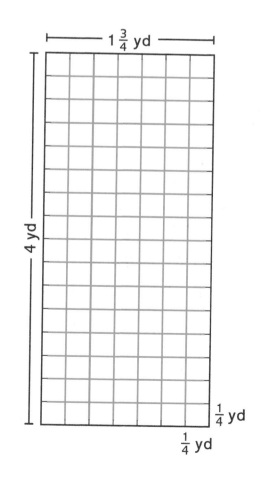

$1\frac{3}{4}$ yd

4 yd

$\frac{1}{4}$ yd

$\frac{1}{4}$ yd

a. How many tiles will Deshawn need to cover the section of the wall? Use the picture to help.

____112____ tiles

b. How many tiles would it take to cover 1 square yard?

____16____ tiles

c. Use your answers to Parts a and b to find the area in square yards of the section of the wall that Deshawn is decorating.

____7____ square yards

____112 / 16 = 7____
(number sentence)

5.NF.4, 5.NF.4b, SMP7

11

Math Boxes

1 Place parentheses in a different place in each problem. Then solve.

$(4 * 8) - 2 =$ ___30___

$4 * (8 - 2) =$ ___24___

SRB
42

2 Find the area of the rectangle.

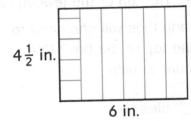

$4\frac{1}{2}$ in.

6 in.

Area = ___27___ square inches

SRB
224-225

3 Write an expression for the number story.

Juno earned 140 points in a game. He lost half of the points and then earned 160 more.   Sample answer:

___$140 - (140 \div 2) + 160$___

SRB
38, 44

4 Complete.

a. 60 inches = ___5___ feet

b. 3 yards = ___9___ feet

c. 4 meters = ___400___ cm

d. 2 miles = ___3,520___ yards

SRB
215-216,
328

5 Expression A: Expression B:
458 − 12   25 + (458 − 12)

Check all that apply.

☑ B is greater than A.

☐ B is twice as much as A.

☑ B is 25 more than A.

SRB
42, 46

6 Write a 5-digit number that has

7 in the ones place,

8 in the hundreds place,

4 in the ten-thousands place,

and 0 in all other places.

_4_ _0_, _8_ _0_ _7_

SRB
66-67

① 5.OA.1 ② 5.NF.4, 5.NF.4b ③ 5.OA.1, 5.OA.2 ④ 5.MD.1
⑤ 5.OA.2 ⑥ 5.NBT.1

Comparing Volume

1 Create Cylinders A and B from half-sheets of paper.
What could you measure about these cylinders?

SRB
230

Sample answers: height; surface area;
distance around the cylinders; weight; volume

2 Which cylinder do you think has a greater volume? Explain your answer.

Answers vary.

3 Test your prediction. Which cylinder has a greater volume? Explain your answer.

Sample answer: Cylinder A has a greater volume than Cylinder
B. I filled A with beans and then poured the beans into B. Not
all of the beans from A fit into B, which means that Cylinder A
holds more and has a greater volume.

4 Think about all the attributes you listed in Problem 1.

a. How are the cylinders different?

Sample answer: They differ in height, in the
distance around them, and in volume.

b. How are the cylinders the same?

Sample answer: The cylinders are both made from the
same-size paper, so they must have the same area on the
outside and the same weight.

5.MD.3, 5.MD.4, SMP3

13

Math Boxes

1 Solve.

   a. $2 * (14 + 6) =$ __40__

   b. $(21 / 3) + 14 =$ __21__

   c. $(10 * 8) - 20 =$ __60__

SRB
42

2 Name three objects that have the attribute of volume.

Answers vary but
should all be
3-dimensional objects.

SRB
230

3 Jo's closet is 6 ft wide and $1\frac{1}{2}$ ft deep. Find the area of the closet floor.

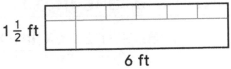

$1\frac{1}{2}$ ft

6 ft

Area = __9__ square feet

SRB
224-225

4 Which numerical expression shows the following calculation? Fill in the circle next to the best answer.

Add seven and three, then multiply by 6.

   Ⓐ $(6 * 7) + 3$

   Ⓑ $(7 + 3) * 6$

   Ⓒ $7 + 3 + 6$

SRB
42, 45

5 **Writing/Reasoning** Explain how you found the area of Jo's closet floor in Problem 3.

Sample answer: The rectangle has six columns that each have $1\frac{1}{2}$ squares, so I added $1\frac{1}{2}$ six times. $1\frac{1}{2} + 1\frac{1}{2} + 1\frac{1}{2} + 1\frac{1}{2} + 1\frac{1}{2} + 1\frac{1}{2} = 6 + 3 = 9$

SRB
224-225

① 5.OA.1 ② 5.MD.3 ③ 5.NF.4, 5.NF.4b
④ 5.OA.2 ⑤ 5.NF.4, 5.NF.4b, SMP2, SMP6

14

Packing Prisms to Measure Volume

1 Use pattern blocks to measure the volume of your rectangular prism. Record your results below. Answers vary.

Our prism has a volume of about _____ **square** pattern blocks.

Our prism has a volume of about _____ **triangle** pattern blocks.

Our prism has a volume of about _____ **hexagon** pattern blocks.

2 What was important to remember as you packed the prism with pattern blocks so you could measure volume as accurately as possible?

Sample answer: We had to pack the blocks as tightly as possible and try to avoid gaps and overlaps.

3 Which pattern block did you need the **most** of to fill your prism? Why?

Sample answer: We used the greatest number of triangles. They are the smallest unit, so we needed the most of them.

4 Which pattern block did you need the **least** of to fill your prism? Why?

Sample answer: We used the least number of hexagons. They are also the largest unit.

5 What other objects could you use to fill the prism?

Sample answers: other pattern blocks, marbles, cubes, dried beans, popcorn, small pebbles

6 What 3-dimensional shape do you think would be easiest to pack tightly into a rectangular prism without gaps or overlaps? Why do you think so?

Sample answer: A cube would pack tightly because each face would fit exactly against the face of another cube. It would also fit in the corners of the prism.

More Areas of Rectangles

Find the area of each rectangle.

1

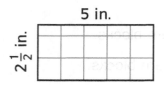

5 in.

$2\frac{1}{2}$ in.

Area = $12\frac{1}{2}$ square inches

2

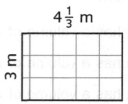

$4\frac{1}{3}$ m

3 m

Area = 13 square meters

3

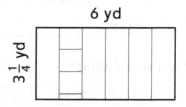

6 yd

$3\frac{1}{4}$ yd

Area = $19\frac{2}{4}$, square yards or $19\frac{1}{2}$

4

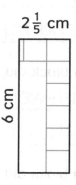

$2\frac{1}{5}$ cm

6 cm

Area = $13\frac{1}{5}$ square centimeters

Try This

5 **a.** What is the area of this rectangle if the sides of the squares are each 1 unit long?

Area = 45 square units

b. What is the area of this rectangle if the sides of the squares are each $\frac{1}{3}$ unit long?

Area = 5 square units

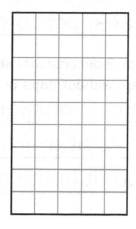

5.NF.4, 5.NF.4b

Math Boxes

1 Box A has a volume of 152 beans. If Box B has a greater volume than Box A, which could be the volume of Box B?

Choose the best answer.

◯ 25 beans

◯ 100 beans

◯ 125 beans

⬭ 200 beans

SRB
230

2 What is the area of a rectangle that is $2\frac{1}{4}$ in. wide and 4 in. long?

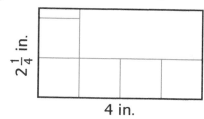

4 in.

Area = ____9____ square inches

SRB
224-225

3 Which expressions are less than $16 - 8$?

Circle ALL that apply.

A. $(16 - 8) * 2$

B. $(16 - 8) - 2$ ⟵ circled

C. $(16 - 8) \div 2$ ⟵ circled

D. $(16 - 8) + 2$

SRB
46

4 Complete.

a. 1 hour = ____60____ minutes

b. $3\frac{1}{2}$ hours = ____210____ minutes

c. $\frac{3}{4}$ hour = ____45____ minutes

d. ____4____ hours = 240 minutes

SRB
215-216,
328

5 Write an expression for each statement.

Add 12 and 8 and multiply the sum by 3.

____$(12 + 8) * 3$____

Subtract 10 from the product of 6 and 8.

____$(6 * 8) - 10$____

SRB
42, 45

6 Josh had 10 goldfish and 2 guppies. Half of the fish are male. Write an expression that shows how many fish are female.

Sample answer:
____$(10 + 2) \div 2$____

SRB
42, 44

■ Math Boxes

① 5.MD.4 ② 5.NF.4, 5.NF.4b ③ 5.OA.2 ④ 5.MD.1 ⑤ 5.OA.1, 5.OA.2 ⑥ 5.OA.1, 5.OA.2

17

Measuring Volume with Cubes

Record your estimates from the Math Message in the second column of the table. Record the actual number of cubes you used to fill the prism in the third column of the table.

Rectangular prism	Estimated Number of Cubes to Fill the Prism	Actual Number of Cubes to Fill the Prism
A	Answers vary.	24
B	Answers vary.	20
C	Answers vary.	18

The cubes in each rectangular prism below are the same size.
Each prism has at least one stack of cubes that goes up to the top.
Find the total number of cubes needed to completely fill each prism.

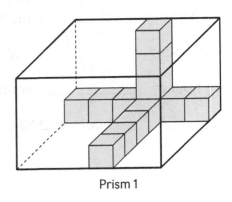

Prism 1

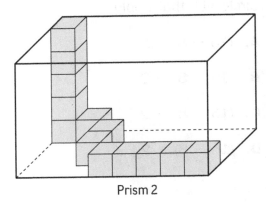

Prism 2

Cubes needed to fill Prism 1: __120__ cubes   Cubes needed to fill Prism 2: __160__ cubes

Volume of Prism 1: __120__ cubic units   Volume of Prism 2: __160__ units³

18   5.MD.3, 5.MD.3a, 5.MD.3b, 5.MD.4, SMP2

Measuring Volume with Cubes (continued)

The cubes in each rectangular prism are the same size.
Each prism has at least one stack of cubes that goes up to the top.
Find the number of cubes needed to completely fill each prism.

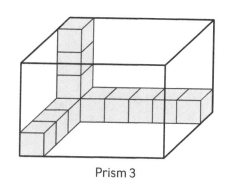

Prism 3

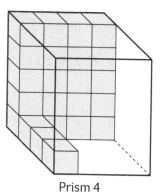

Prism 4

Cubes needed to fill Prism 3: ____96____ cubes

Volume of Prism 3: ____96____ cubic units

Cubes needed to fill Prism 4: ____80____ cubes

Volume of Prism 4: ____80____ units³

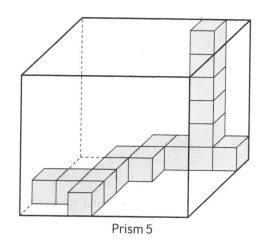

Prism 5

Try This

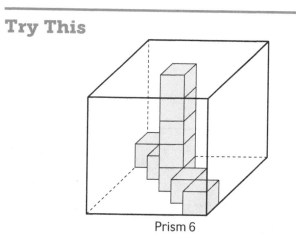

Prism 6

Cubes needed to fill Prism 5: ____210____ cubes

Volume of Prism 5: ____210____ units³

Cubes needed to fill Prism 6: ____125____ cubes

Volume of Prism 6: ____125____ cubic units

5.MD.3, 5.MD.3a, 5.MD.3b, 5.MD.4, SMP2

Math Boxes

1 Solve.

a. $(4 * 3) + (9 / 3) =$ __15__

b. $4 * (2 + 7) - 6 =$ __30__

c. __3__ $= (10 + 8) / (54 / 9)$

SRB
42

2 If you want to know how many boxes of markers will fit in a drawer, do you need to know the length, area, or volume of the drawer?

__Volume__

SRB
218-219,
221, 230

3 Find the area of a table that is 3 feet wide and $2\frac{1}{2}$ feet long.

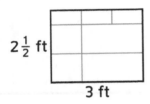

$2\frac{1}{2}$ ft

3 ft

Area = __$7\frac{1}{2}$__ square feet

SRB
224-225

4 Write the following expressions using numbers.

a. seven times the sum of 6 and 4

__$7 * (6 + 4)$__

b. the difference of sixteen and eight, divided by two

__$(16 - 8) \div 2$__

SRB
42, 45

5 **Writing/Reasoning** How did you decide which type of measurement you would need to know in Problem 2?

Sample answer: If I want to know about filling the space of something 3-dimensional, I need to find the volume.

SRB
218-219,
221, 230

① 5.OA.1 ② 5.MD.3 ③ 5.NF.4, 5.NF.4b ④ 5.OA.1, 5.OA.2

20   ⑤ 5.MD.3, SMP6

Converting Measurements

Solve. If necessary, look up measurement equivalents in the *Student Reference Book*.

1 Record measurement equivalents in the 2-column tables below.

Minutes	Seconds
5	300
10	600
15	900
20	1,200
25	1,500

Kiloliters	Liters
5	5,000
$1\frac{1}{2}$	1,500
$2\frac{1}{2}$	2,500
4	4,000
$3\frac{1}{2}$	3,500

Miles	Yards
1	1,760
2	3,520
3	5,280
4	7,040
5	8,800

2 Students recorded their running distances over the weekend using different units. Complete the table to convert them to the same units.

	Kilometers	Meters
Jason	3	3,000
Kayla	$4\frac{1}{2}$	4,500
Lohan	$2\frac{1}{2}$	2,500
Malik	5	5,000
Jada	$3\frac{1}{2}$	3,500

3 Jordan needed to convert measurements for his recipes. Fill in the blanks.

2 quarts milk = __8__ cups

32 oz flour = __2__ lb

8 cups rice = __4__ pints

__4__ cups pasta = 2 pints

$2\frac{1}{2}$ cups water = __20__ fl oz

$\frac{1}{2}$ cup oil = __8__ tbs

4 Mahalia is making cloth napkins. She bought fabric that is 12 inches wide and 6 yards long.

How many napkins can Mahalia make if each napkin is 1 foot by 1 foot? __18__ napkins

5.MD.1

21

Cube-Stacking Problems Using Layers

SRB
232

Complete the table for each rectangular prism.

Rectangular Prism	Number of Cubes in 1 Layer	Number of Layers	Total Number of Cubes That Fill the Prism	Volume of the Prism
G	10	5	50	50 cubic units
H	16	8	128	128 cubic units
I	20	6	120	120 cubic units
J	15	7	105	105 cubic units
K	25	5	125	125 cubic units
L	20	7	140	140 cubic units

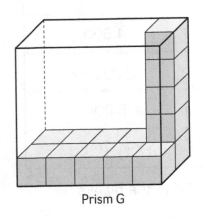

Prism G

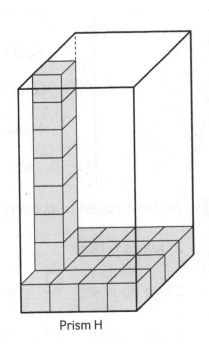

Prism H

5.MD.3, 5.MD.3a, 5.MD.3b, 5.MD.4

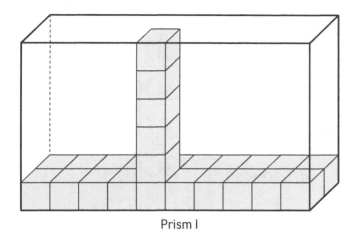

Prism I

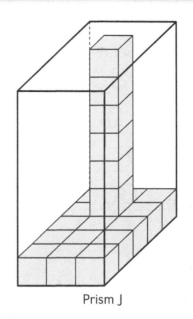

Prism J

Try This

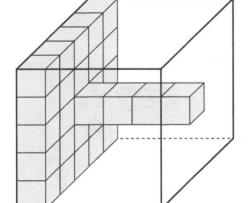

Prism K

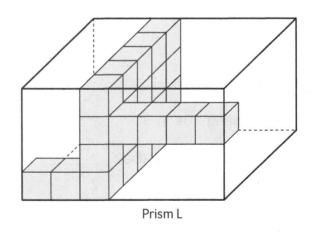

Prism L

5.MD.3, 5.MD.3a, 5.MD.3b, 5.MD.4   23

1 What is the value of the 8 in the following numbers?

a. 1,384 ___80___

b. 8,294 ___8,000___

c. 418 ___8___

d. 6,897 ___800___

SRB
66-67

2 Solve.

a. 3 * 10 = ___30___

b. 3 * 100 = ___300___

c. 3 * 1,000 = ___3,000___

d. 30 * 10 = ___300___

SRB
95

3 Write four multiples of 4.

Sample answers:

___24___ , ___16___ , ___40___ , ___28___

SRB
72

4 Solve.

a.
```
    2 3
  *   3
  ─────
    6 9
```

b.
```
  2 4 2
*     2
───────
  4 8 4
```

SRB
100-101,
104

5 Solve.

How many 10s in 30? ___3___

How many 10s in 300? ___30___

How many 7s in 21? ___3___

How many 7s in 210? ___30___

SRB
106

6 Solve.

50 + 6 = ___56___

300 + 20 = ___320___

200 + 50 + 6 = ___256___

SRB
70

① 5.NBT.1 ② 5.NBT.2 ③ 5.NBT.6 ④ 5.NBT.5 ⑤ 5.NBT.6
⑥ 5.NBT.1

24

Math Boxes

Finding Volume Using Formulas

Use a formula to find the volume of each prism. Record the formula you used.

SRB
233

1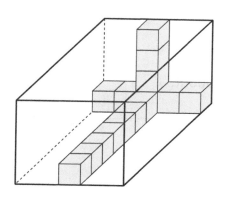

Volume: _____ 160 units³ _____

Formula: _____ V = l × w × h,
or V = 8 × 5 × 4 _____

2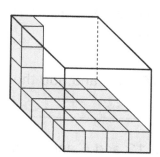

Volume: _____ 80 units³ _____

Formula: _____ V = B × h,
or V = 20 × 4 _____

3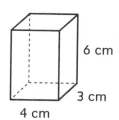

6 cm
3 cm
4 cm

Volume: _____ 72 cm³ _____

Formula: _____ V = l × w × h,
or V = 4 × 3 × 6 _____

4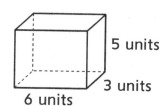

5 units
3 units
6 units

Volume: _____ 90 units³ _____

Formula: _____ V = l × w × h,
or V = 6 × 3 × 5 _____

5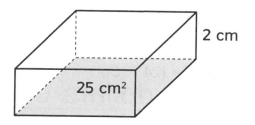

2 cm
25 cm²

Volume: _____ 50 cm³ _____

Formula: _____ V = B × h,
or V = 25 × 2 _____

Try This

6 A rectangular prism has a volume of 36 cubic units. Write two different possible sets of dimensions for the prism.
Sample answers:

Set 1:

length = __ 2 units __

width = __ 9 units __

height = __ 2 units __

Set 2:

length = __ 3 units __

width = __ 3 units __

height = __ 4 units __

5.MD.5, 5.MD.5a, 5.MD.5b

25

Writing and Interpreting Expressions

Write an expression that models the calculation described in words.

1 The sum of 13 and 12, which is then multiplied by 2

$(13 + 12) * 2$

2 Divide 16 by 4 and add the sum of 3 and 8 to the quotient.

$(16 / 4) + (3 + 8)$

3 Multiply 12 and 6 and divide the product by 9.

$12 * 6 / 9$, or $(12 * 6) / 9$

Without calculating, circle the expression with the greater value.

4 $3 * (126 + 12)$ $\boxed{6 * (126 + 12)}$

5 $\boxed{(18 - 8) / 2}$ $(18 - 8) / 5$

6 Explain how you knew which expression had a greater value in Problem 5.

Sample answer: If I divide by 5, that creates more equal parts than if I divide by 2, so I know the parts will be smaller. Dividing by 2 will give a larger quotient.

7 Ivan was playing a video game. He had 1,300 points and on the next level earned 120 more. Then he lost 12 points. When his turn ended, his score doubled. Write an expression that shows the number of points Ivan has at the end of his turn.

$(1,300 + 120 - 12) * 2$

8 Write a situation that can be modeled by the expression $6 * (24 - 5)$.

Sample answer: Sam earns 24 dollars each week at his job. He keeps 5 dollars and puts the rest in his savings account. After 6 weeks, how much money does Sam have in his savings account?

5.OA.1, 5.OA.2, SMP4

Math Boxes

1 What is the volume of the prism?

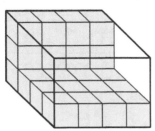

Volume = __48__ cubic units

SRB
231-232

2 Raoul says he can find the area of a rectangle that is $1\frac{1}{2}$ in. by 3 in. using addition. Write a number model that Raoul could use to find the area of the rectangle.

Area = $\dfrac{1\frac{1}{2} + 1\frac{1}{2} + 1\frac{1}{2}}{}$

SRB
224-225

3 This prism is made of unit cubes. Use $V = B * h$ to find the volume.

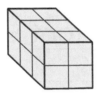

Volume = $\underset{\substack{\text{area of} \\ \text{base}}}{6} * \underset{\text{height}}{2} = 12$ cubic units

SRB
231-233

4 Insert parentheses to make the equations true.

a. $(2 + 3) * 4 = 20$

b. $4 * (5 - 3) = 8$

SRB
42

5 **Writing/Reasoning** How did you figure out the volume of the prism in Problem 1?

Sample answer: I counted the cubes in the bottom layer and got 16. I knew that 3 layers of cubes would fill the prism, so I added 16 three times and got 48.

SRB
231-232

① 5.MD.3, 5.MD.3a, 5.MD.3b, 5.MD.4 ② 5.NF.4, 5.NF.4b
③ 5.MD.5, 5.MD.5a, 5.MD.5b ④ 5.OA.1 ⑤ 5.MD.3, 5.MD.3a,
5.MD.3b, 5.MD.4, SMP1, SMP6

Converting Cubic Units

1. Is a cubic inch larger or smaller than a cubic centimeter? How do you know?

   SRB 235

   Sample answer: Larger. An inch is longer than a centimeter, so a cube with 1-inch edges is larger than a cube with 1-centimeter edges.

2. List objects with volumes you might measure in cubic inches.

   Sample answers: a shoe box, a desk drawer

3. a. How many cubic inches do you think are in a cubic foot? Answers vary.

   b. How many inches are in a foot? 12

   c. How many square inches are in a square foot?

   144 square inches

   How did you find your answer?

   Since 12 in. = 1 ft, a square foot would be 12 in. long and 12 in. high. 12 × 12 = 144

   d. How many cubic inches are in a cubic foot?

   1,728 cubic inches

   How did you find your answer? There are 144 square inches in 1 square foot, so it takes 144 cubic inches to cover a square foot. A cubic foot is 12 inches high, so there would be 12 layers. 144 × 12 = 1,728

4. List objects with volumes you might measure in cubic feet.

   Answers vary, but objects should be larger than those listed for cubic inches, such as a closet, a locker, or the trunk of a car.

5.MD.1, 5.MD.3, 5.MD.3a, 5.MD.3b, 5.MD.4, SMP6

5 How many cubic feet are in a cubic yard?

___27___ cubic feet

How did you find your answer? Sample answer: I know that there are 3 feet in a yard. That means that there are 3 × 3, or 9 square feet, in a square yard and 9 × 3, or 27 cubic feet, in a cubic yard.

6 List objects with volumes you might measure in cubic yards. Answers vary, but objects should be larger than those listed for cubic feet, such as a warehouse, a hallway, or a cargo ship.

7 Deena's family has a freezer that is 2 yards in width, 1 yard in length, and 1 yard in height.

a. What is the volume of the freezer?

___2___ cubic yards

b. How many cubic feet of food will fit in the freezer?

___54___ cubic feet

How did you find your answer? Sample answer: I converted the dimensions to feet. 2 yards = 6 feet, and 1 yard = 3 feet. 6 × 3 × 3 = 54 cubic feet

c. Do you think cubic yards or cubic feet are better units to measure the volume of the freezer? Why?

Answers vary.

More Cube-Stacking Problems

The cubes in each rectangular prism are the same size. Find the volume of each prism.

1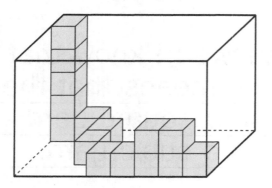

Volume = __160__ cubic units

2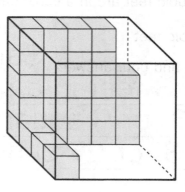

Volume = __100__ cubic units

3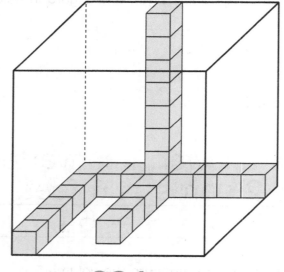

Volume = __384__ cubic units

Try This

4 The stack of cubes represents the height of a rectangular prism.

Draw the outline of a rectangular base on the grid paper so that the volume of the prism would be 60 cubic units.

Sample rectangle:

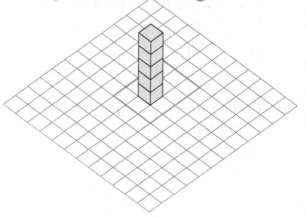

What is the area of the base you drew?

__12__ square units

5.MD.3, 5.MD.3a, 5.MD.3b, 5.MD.4, SMP2

Math Boxes

1 Can A has a volume of 25 beans. Can B holds twice as many beans as Can A. What is the volume of Can B?

Volume = ___50___ beans

SRB
230

2 Find the area of a rectangle that is $1\frac{1}{2}$ feet by 2 feet.

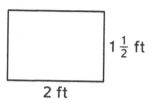

$1\frac{1}{2}$ ft

2 ft

Area = ___3___ square feet

SRB
224-225

3 If 329 + 671 = 1,000, what is 2 * (329 + 671)?

___2,000___

SRB
46

4 Complete.

a. 100 yd = ___300___ ft

b. 2 mi = ___10,560___ ft

c. ___72___ in. = 2 yd

d. ___$2\frac{1}{2}$___ ft = 30 in.

SRB
215-216,
328

5 Solve.

a. (14 + 2) / 8 = ___2___

b. (36 / 6) + (42 / 7) = ___12___

c. 3 * (50 + 20 + 30) = ___300___

SRB
42

6 Maria bought 5 tickets for $20 each. Each ticket had a $2 fee. Write an expression that shows how much Maria paid for the tickets.

Sample answer:
(20 + 2) * 5

SRB
42, 44

① 5.MD.4 ② 5.NF4, 5.NF.4b ③ 5.OA.2 ④ 5.MD.1
⑤ 5.OA.1 ⑥ 5.OA.1, 5.OA.2

31

Estimating Volumes of Instrument Cases

In Problems 1–3, use the mathematical models to estimate the volumes of the instrument cases.

1 Trombone case

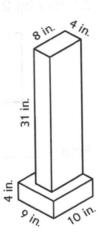

The volume of the trombone case is about ___1,352___ in.³.

2 French horn case

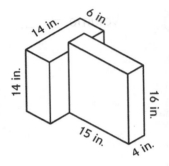

The volume of the French horn case is about ___2,136___ in.³.

3 Xylophone case

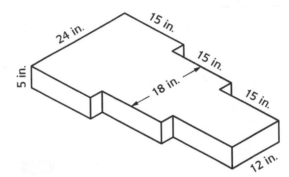

The volume of the xylophone case is about ___4,050___ in.³.

Try This

4 Asher needs to take the xylophone, the trombone, and the French horn with him to a band concert. His trunk has 13 cubic feet of cargo space. Can he fit all three cases in his trunk? Explain how you know. Sample answer: Yes, The three instruments together take up 7,538 in.³ of space. I know there are 1,728 in.³ in 1 ft³, so 7,538 in.³ is the same as 7,538 / 1,728, or about 4 ft³. Asher has 13 ft³ in his trunk, so the cases should fit.

32　5.MD.1, 5.MD.5, 5.MD.5b, 5.MD.5c, SMP4

Math Boxes

① What is the volume of the prism?

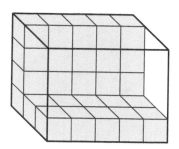

Volume = _____60_____ cubic units

② A doormat is 3 feet by $2\frac{1}{2}$ feet. What is the area of the doormat?

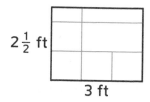

$2\frac{1}{2}$ ft

3 ft

Area = _____$7\frac{1}{2}$_____ ft²

③ This prism is made of unit cubes. Use $V = B \times h$ to find the volume.

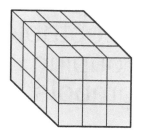

Volume = $\underset{\substack{\text{area of}\\\text{the base}}}{12} \times \underset{\text{height}}{3} = 36$ cubic units

④ Insert parentheses to make the equations true.

a. $5 * (4 - 2) = 10$

b. $(8 - 7) * 25 = 25$

c. $36 \div (6 - 5) = 36$

⑤ Writing/Reasoning What does the area of the base tell you about the number of cubes that fit in the prism in Problem 3?

Sample answer: If I know the area of the base, I can figure out how many unit cubes are in one layer. Then I can multiply by the number of layers to get the volume.

① 5.MD.3, 5.MD.3a, 5.MD.3b, 5.MD.4 ② 5.NF.4, 5.NF.4b
③ 5.MD.5, 5.MD.5a, 5.MD.5b ④ 5.OA.1 ⑤ 5.MD.5, 5.MD.5a,
5.MD.5b, SMP2, SMP7

Math Boxes

Understanding Grouping Symbols

Evaluate the following expressions.

1. $10 * [13 + (12 - 7)] = $ __180__

2. __8__ $= \{(5 * 6) + 2\} / 4$

3. __45__ $= \{13 + (2 * 1)\} * 3$

4. $64 / [20 - (4 * 3)] = $ __8__

Insert grouping symbols to make the following number sentences true. **Brackets, braces, and parentheses are all acceptable grouping symbols.**

5. $4 = 4 * [6 - (2 + 3)]$

6. $300 \div \{(6 + 4) * (2 + 8)\} = 3$

7. $70 / [13 - (2 + 1)] = 7$

8. $160 = 8 * [16 + \{(12 - 4) \div 2\}]$

Write an expression that models the story. Then evaluate the expression.

9. Tommy had a bag of 100 balloons. He took out 2 red balloons and 1 blue balloon for each party favor. He created 12 party favors. How many balloons did Tommy have left?

   Expression: __$100 - [12 * (2 + 1)]$__

   Answer: __64__ balloons

10. A grocery store received a shipment of 100 cases of apple juice. Each case contained four 6-packs of cans. After inspection, the store found that 9 cans were damaged. How many cans were undamaged?

    Expression: __$\{100 * (4 * 6)\} - 9$__

    Answer: __2,391__ cans

5.OA.1, 5.OA.2, SMP4

Math Boxes

1 Jayden filled a tub with boxes of markers. He put 2 boxes in each layer. It took 4 layers to fill the tub. What is the volume of the tub?

Volume = _____8_____ marker boxes

SRB
230, 232

2 Imani's room is 3 yards by $4\frac{1}{4}$ yards. She is getting new carpet.

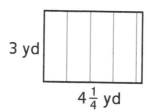

3 yd

$4\frac{1}{4}$ yd

Is 13 square yards of carpet enough to cover her room? __Yes.__

SRB
224-225

3 Write an expression with a value that is twice as much as 236 + 912.

Sample answer:
2 * (236 + 912)

SRB
45-46

4 Complete.

a. 1.5 km = __1,500__ m

b. 36 in. = __1__ yd

c. 3 m = __3,000__ mm

d. 5 dm = __50__ cm

SRB
215-216, 328

5 Solve.

a. $\left(\frac{1}{2} + \frac{1}{2}\right) * 5 =$ __5__

b. $10 * (300 \div 30) =$ __100__

c. $6 * \left(1\frac{1}{2} + 1\frac{1}{2}\right) =$ __18__

SRB
42,
186-187

6 Milo has 3 dogs and 2 cats. Each pet eats 2 cups of food a day.

Which expression(s) below show how much food Milo's pets eat in one day?

Fill in the circle next to all that apply.

Ⓐ (3 * 2) + 2

🅱 (3 + 2) * 2

Ⓒ 3 + (2 + 2)

🅳 (3 * 2) + (2 * 2)

SRB
42, 44

① 5.MD.4 ② 5.NF.4, 5.NF.4b ③ 5.OA.2 ④ 5.MD.1
⑤ 5.OA.1 ⑥ 5.OA.1, 5.OA.2

35

Math Boxes: Preview for Unit 2

Math Boxes

① Use the numbers below to solve. Use each digit once.

7, 1, 0, 2, 9

a. Write the largest number you can.

_97,210_

b. Write the smallest number you can. (Do not start with 0.)

_10,279_

SRB
66-67

② Solve.

a. 6 * 100 = _600_

b. 6 * 10 = _60_

c. 6 * 1,000 = _6,000_

d. 60 * 10 = _600_

SRB
95

③ Which of the following are all multiples of 6? Choose the best answer.

◯ 24, 72, 19, 36

◯ 12, 27, 42, 18

⬤ 60, 48, 12, 24

◯ 6, 56, 42, 30

SRB
72

④ Solve.

a.
```
    5 6
  *   4
  2 2 4
```

b.
```
    4 2 3
  *     6
  2,5 3 8
```

SRB
100-101, 104

⑤ Solve.

27 ÷ 9 = _3_

270 ÷ 9 = _30_

_6_ = 42 ÷ 7

_60_ = 420 ÷ 7

SRB
106

⑥ Write the following number in expanded form:

23,465 = _Sample answer:_

_20,000 + 3,000 + 400 + 60 + 5_

SRB
70

① 5.NBT.1 ② 5.NBT.2 ③ 5.NBT.6 ④ 5.NBT.5
36 ⑤ 5.NBT.6 ⑥ 5.NBT.1

Place-Value Relationships

Math Message

SRB
66-67,
70

1 What is the value of the 2 in the following numbers?

2 _____2_____

23 _____20_____

230 _____200_____

2,300 _____2,000_____

23,000 _____20,000_____

2 What is the value of the 6 in the following numbers?

65,000 _____60,000_____

6,500 _____6,000_____

650 _____600_____

65 _____60_____

6 _____6_____

---

3 Write these numbers in expanded form. Sample answers:

a. $2,387,926 = $ (2 * 1,000,000) + (3 * 100,000) +
(8 * 10,000) + (7 * 1,000) + (9 * 100) + (2 * 10) + (6 * 1)

b. $92,409,224 = $ 90,000,000 + 2,000,000 + 400,000 +
9,000 + 200 + 20 + 4

4 Write these numbers in standard notation.

a. 4 [100,000s] + 5 [10,000s] + 0 [1,000s] + 3 [100s] + 6 [10s] + 2 [1s] = 450,362

b. (9 * 10,000) + (3 * 1,000) + (4 * 100) + (9 * 10) + (1 * 1) = 93,491

c. 3 ten-thousands + 2 thousands + 5 hundreds + 7 tens + 9 ones = 32,579

5 How does expanded form help you see the patterns in our place-value system?

Sample answer: Expanded form shows how much each digit is worth. It shows how the value of each digits depends on its place in the number.

6 Write the value of the 4 in each number.

a. 348,621 _40,000_

b. 24,321 _4,000_

c. 624,876,712 _4,000,000_

d. 13,462 _400_

e. 463,295 _400,000_

f. 942 _40_

7 Write the value of the identified digit. Then fill in the blank with "10 times" or "$\frac{1}{10}$ of."

a. What is the value of 7 in 732? _700_   In 7,328? _7,000_

The value of the 7 in 732 is _$\frac{1}{10}$ of_ the value of 7 in 7,328.

b. What is the value of 4 in 32,940? _40_   In 32,904? _4_

The value of 4 in 32,940 is _10 times_ the value of 4 in 32,904.

c. What is the value of 2 in 30,275? _200_   In 18,921? _20_

The value of 2 in 30,275 is _10 times_ the value of 2 in 18,921.

d. What is the value of 1 in 90,106? _100_   In 21,000? _1,000_

The value of the 1 in 90,106 is _$\frac{1}{10}$ of_ the value of 1 in 21,000.

8 a. Write a number where 5 is worth 500. _Sample answer: 2,542_

b. Write a number where 5 is worth 10 times as much as the number you wrote in Part a.
Sample answer: 5,242

c. How did the position of the 5 change in your number in Part b?
The 5 moved one place to the left.

9 a. Write a number where 3 is worth 30,000 and 2 is worth 20.
Sample answer: 34,129

b. Write a number where 3 is worth $\frac{1}{10}$ as much and 2 is worth 10 times as much as
the number you wrote in Part a. _Sample answer: 3,200_

c. How did the position of the 3 change in your number in Part b?
The 3 moved one place to the right.

Finding Volumes of Rectangular Prisms

The cubes in each rectangular prism are the same size. Each prism has at least one stack of cubes that goes up to the top. Find the number of cubes needed to completely fill each prism. Then use the formula to help you write a number sentence that represents the volume.

 1

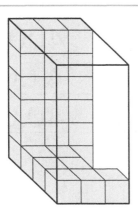

___72___ cubes to fill the prism

$V = B * h$

__72__ = __12__ * __6__

 2

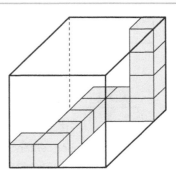

___80___ cubes to fill the prism

$V = l * w * h$

__80__ = __4__ * __5__ * __4__

 3

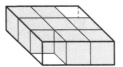

___9___ cubes to fill the prism

$V = l * w * h$

__9__ = __3__ * __3__ * __1__

 4

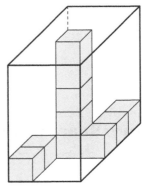

___75___ cubes to fill the prism

$V = B * h$

__75__ = __15__ * __5__

5.MD.3, 5.MD.3a, 5.MD.3b, 5.MD.4,
5.MD.5, 5.MD.5a, SMP2

Math Boxes

1 Find the area of the rectangle.
Write a number sentence.

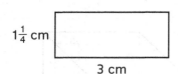

$1\frac{1}{4}$ cm

3 cm

Area = $3\frac{3}{4}$ cm²

$1\frac{1}{4} + 1\frac{1}{4} + 1\frac{1}{4} = 3\frac{3}{4}$

(number sentence)

Sample number sentence.

SRB
224-225

2 What is the value of the 4 in the following numbers?

a. 42 ___40___

b. 420 ___400___

c. 4,200 ___4,000___

d. 42,000 ___40,000___

SRB
66-67

3 Complete.

a. 6 feet = ___72___ inches

b. ___2___ tons = 4,000 pounds

c. $\frac{1}{2}$ pound = ___8___ ounces

d. 9 yards = ___27___ feet

SRB
215-217,
328

4 Find the volume of the prism.

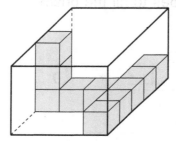

Volume = ___60___ cubic units

SRB
231-232

5 **Writing/Reasoning** How did you find the area of the rectangle in Problem 1?

Sample answer: I thought about covering the rectangle
with squares and partial squares. 3 squares would go
across a row and $1\frac{1}{4}$ squares would go up a column.
I added $1\frac{1}{4}$ three times because there were 3 columns
of $1\frac{1}{4}$ squares. $1\frac{1}{4} + 1\frac{1}{4} + 1\frac{1}{4} = 3\frac{3}{4}$

SRB
224-225

① 5.NF.4, 5.NF.4b ② 5.NBT.1 ③ 5.MD.1 ④ 5.MD.3,
40   5.MD.3a, 5.MD.3b, 5.MD.4 ⑤ 5.NF.4, 5.NF.4b, SMP6

Powers of 10 and Exponential Notation

Follow your teacher's instructions to complete this table.

Exponential Notation	Product of 10s	Standard Notation
10^6	10 * 10 * 10 * 10 * 10 * 10	1,000,000
10^5	10 * 10 * 10 * 10 * 10	100,000
10^8	10 * 10 * 10 * 10 * 10 * 10 * 10 * 10	100,000,000
10^3	10 * 10 * 10	1,000

Complete the number sentences.

1. $3 * 10^2 = 3 * 100 =$ __300__

2. $7 * 10^9 = 7 *$ __1,000,000,000__ $=$ __7,000,000,000__

3. $3 * 10^4 =$ __3__ $*$ __10,000__ $=$ __30,000__

4. $25 * 10^3 =$ __25__ $*$ __1,000__ $=$ __25,000__

5. __93 * 10^6__ $= 93 *$ __1,000,000__ $= 93,000,000$

6. The numbers in Problems 1–5 are the answers to the following questions. Fill in each blank with your best guess. Write your answer in exponential or standard notation.

   a. The distance from the sun to Earth is about __$93 * 10^6$, or 93,000,000__ miles.

   b. The Statue of Liberty is about __$3 * 10^2$, or 300__ feet tall.

   c. In 2014, the population of Earth was about __$7 * 10^9$, or 7,000,000,000__ people.

   d. If you walked around Earth's equator, you would walk about __$25 * 10^3$, or 25,000__ miles.

   e. Mount Everest, the highest mountain on Earth, is about __$3 * 10^4$, or 30,000__ feet high.

Expanded Form with Powers of 10

Numbers in expanded form are written as addition expressions showing the value of each digit.

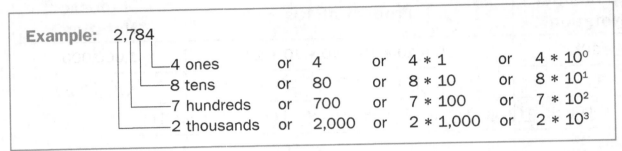

Example: 2,784

4 ones	or	4	or	$4 * 1$	or	$4 * 10^0$
8 tens	or	80	or	$8 * 10$	or	$8 * 10^1$
7 hundreds	or	700	or	$7 * 100$	or	$7 * 10^2$
2 thousands	or	2,000	or	$2 * 1,000$	or	$2 * 10^3$

2,784 can be written in expanded form in different ways.

- As an addition expression: $2,000 + 700 + 80 + 4$

- As the sum of multiplication expressions involving powers of 10: $(2 * 1,000) + (7 * 100) + (8 * 10) + (4 * 1)$

- As the sum of multiplication expressions using exponents to show the powers of 10:
  $(2 * 10^3) + (7 * 10^2) + (8 * 10^1) + (4 * 10^0)$

1 **a.** Write 6,125 in expanded form as an addition expression.

$6,000 + 100 + 20 + 5$

b. Write 6,125 in expanded form as the sum of multiplication expressions involving powers of 10.

$(6 * 1,000) + (1 * 100) + (2 * 10) + (5 * 1)$

c. Write 6,125 in expanded form as the sum of multiplication expressions using exponents to show the powers of 10.

$(6 * 10^3) + (1 * 10^2) + (2 * 10^1) + (5 * 10^0)$

2 Write each number in standard notation.

a. $12 * 10^5$ 1,200,000   **b.** $4 * 10^8$ 400,000,000

3 Write each number using exponential notation and powers of 10.

a. 30,000 $3 * 10^4$ **b.** 4,200,000 $42 * 10^5$

Solving a Real-World Volume Problem

Josef and his mother are renting a storage unit. They want to rent the largest unit. The dimensions of the available units are shown below. Calculate the volume of each unit. Write the formula you use and show your work. Circle the storage unit you think they should rent.

Storage Unit 1

8 ft
6 ft
5 ft

Formula: V = ___B * h, or l * w * h___

The volume of Storage Unit 1 is ___240___ ft³.

Storage Unit 2

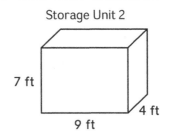

7 ft
4 ft
9 ft

Formula: V = ___B * h, or l * w * h___

The volume of Storage Unit 2 is ___252___ ft³.

Storage Unit 3

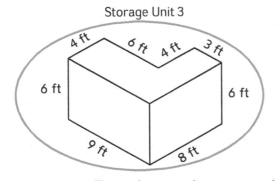

4 ft   6 ft   4 ft   3 ft
6 ft   6 ft
9 ft   8 ft

Formula: V = ___B * h, or l * w * h___

The volume of Storage Unit 3 is ___288___ ft³.

Storage Unit 4

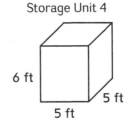

6 ft
5 ft
5 ft

Formula: V = ___B * h, or l * w * h___

The volume of Storage Unit 4 is ___150___ ft³.

5.MD.5, 5.MD.5a, 5.MD.5b, 5.MD.5c, SMP4

43

Math Boxes

1 Solve.

a. $(24 ÷ 8) × 4 =$ ___12___

b. $4 + (15 / 3) =$ ___9___

c. $[(6 + 4) × 3] + 6 =$ ___36___

d. $4 × \{5 + (10 ÷ 2)\} =$ ___40___

SRB
42-43

2 Write a 4-digit number with
4 in the hundreds place,
8 in the thousands place,
3 in the ones place,
and 7 in the tens place.

__8__ , __4__ __7__ __3__

SRB
66-67

3 Write 23,436 in expanded form.

Sample answer:
2 × 10,000 + 3 ×
1,000 + 4 × 100 +
3 × 10 + 6 × 1

SRB
70

4 Find the volume of the prism. Use the
formula: $V = l × w × h$.

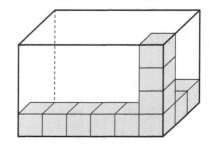

Volume = ___6___ × ___3___ × ___4___ = ___72___ units³

SRB
231-233

5 Write each power of 10 in exponential
notation.

a. $10 × 10 × 10 =$ ___10^3___

b. $10 × 10 × 10 × 10 × 10 =$ ___10^5___

c. $10 × 10 × 10 × 10 × 10 × 10 × 10 =$
___10^7___

d. $10 × 10 × 10 × 10 × 10 × 10 × 10 ×$
$10 × 10 =$ ___10^9___

SRB
68

6 Jonah's sister is 10 years old. Jonah is
8 years younger than twice his sister's
age. Write an expression for Jonah's age.

Sample answer:
$(2 × 10) − 8$

SRB
42, 44

① 5.OA.1 ② 5.NBT.1 ③ 5.NBT.1 ④ 5.MD.5, 5.MD.5a,
44 5.MD.5b ⑤ 5.NBT.2 ⑥ 5.OA.2

Estimating with Powers of 10

Use estimation to solve.

SRB
83,
97-98

1   A hardware store sells ladders that extend up to 12 feet. The store's advertising says:

Largest inventory in the country! If you put all our ladders end to end, you could climb to the top of the Empire State Building!

The company has 295 ladders in stock. The Empire State Building is 1,453 feet tall.

Is it true that the ladders would reach the top of the building? __Yes.__

Explain how you solved the problem. Sample answer: To find out exactly how high all the ladders would reach, I would multiply 12 * 295. To estimate, I multiplied 10 * 300. That's 3,000 feet, enough to get to the top of a 1,453-foot building.

2   The school library received a donation of 42 boxes of books. Each box contains 15–18 books. The library has 10 empty bookshelves. Each bookshelf can hold up to 60 books.

Does the library have enough shelf space for all the new books? __No.__

Explain how you solved the problem. Sample answer: I rounded 42 to 40 and 15–18 to 20 and multiplied 40 * 20. About 800 books were donated. I multiplied 10 * 60 to find that there's only room for 600 books. There is not enough space for the new books.

3   Nishant is in charge of collecting cereal box tops to trade in for technology items for his school. He keeps the box tops in 38 folders. Each folder contains 80 box tops.

Does Nishant have enough box tops to

trade in for a printer? __Yes.__

For a digital camera? __No.__

For a tablet computer? __No.__

Item	Number of Box Tops Needed
Printer	1,500
Digital camera	3,500
Tablet computer	5,000

Explain how you solved the problem. Sample answer: Nishant has 38 * 80 box tops. To estimate, I rounded 38 to 40, then multiplied 40 * 80. Nishant has about 3,200 box tops. That's enough to get a printer, but not a camera or a computer.

5.NBT.2, SMP6

45

Writing and Comparing Expressions

1. Write each statement as an expression using grouping symbols. Do not evaluate the expressions.

   a. Find the sum of 13 and 7, then subtract 5. $(13 + 7) - 5$

   b. Multiply 3 and 4 and divide the product by 6. $(3 \times 4) \div 6$

   c. Add 138 and 127 and multiply the sum by 5. $(138 + 127) \times 5$

   d. Divide the sum of 45 and 35 by 10. $(45 + 35) \div 10$

   e. Add 300 to the difference of 926 and 452. $(926 - 452) + 300$

2. Compare the two expressions. Do not evaluate them. How are their values different?

   a. 2 * (489 + 126) and 489 + 126

   The value of $2 * (489 + 126)$ is twice as large as the value of $489 + 126$.

   b. (367 × 42) − 328 and 367 × 42

   The value of $(367 \times 42) - 328$ is 328 less than the value of 367×42.

3. Below are advertisements for two stores having sales on T-shirts.

   Sample answers:

   a. Write an expression that represents the cost of two shirts at each store.

Shirts-R-Us	T-Shirt Mart
Half off T-shirts! Regular price: $14.00.	T-shirt Sale Price: $8.00.
Expression: $\left(\frac{1}{2} * \$14.00\right) + \left(\frac{1}{2} * \$14.00\right)$	Expression: $\$8.00 + \$8.00$

   b. Which store has the better deal? How do you know?

   Shirts-R-Us; $\frac{1}{2}$ of $14.00 is less than $8.00, so the shirts will cost less at Shirts-R-Us.

5.OA.1, 5.OA.2, SMP1, SMP4

Math Boxes

1 Krista has a garden that is 5 yards by $2\frac{1}{2}$ yards and wants to cover it with compost. The compost is sold in bags that cover 5, 15, or 30 square yards. Which size bag should Krista buy?

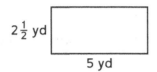

$2\frac{1}{2}$ yd

5 yd

Fill in the circle next to the best answer.

(A) 5 square yards

(B) 15 square yards

(C) 30 square yards

SRB
224-225

2 What is the value of the 8 in the following numbers?

a. 38 ___8___

b. 382 ___80___

c. 832 ___800___

d. 8,432 ___8,000___

e. 85,432 ___80,000___

SRB
66-67

3 Complete.

a. 5 m = ___5,000___ mm

b. ___8___ kg = 8,000 g

c. 10 liters = ___10,000___ milliliters

d. 8 dm = ___80___ cm

e. 5 metric tons = ___5,000___ kg

SRB
215-217,
328

4 Find the volume of the prism.

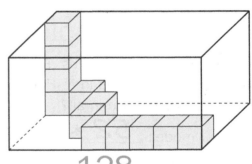

Volume = ___128___ cubic units

SRB
231-232

5 **Writing/Reasoning** Explain what happens to the value of the 8 as it moves one place to the left in Problem 2.

Sample answer: Each time the 8 moves one place to the left, it is worth 10 times as much.

SRB
66-67

① 5.NF.4, 5.NF.4b ② 5.NBT.1 ③ 5.MD.1 ④ 5.MD.3,
5.MD.3a, 5.MD.3b, 5.MD.4 ⑤ 5.NBT.1, SMP7

47

Practicing Multiplication Strategies

For Problems 1 and 2, solve one problem using U.S. traditional multiplication. Solve the other problem using any strategy. Show your work. Be sure to check that each answer makes sense.

SRB
44, 100, 102, 104

1
```
    23
  *  6
  138
```

2
```
    76
  *  5
  380
```

3 Choose Problem 1 or Problem 2. Explain how you checked to see whether your answer made sense.

Sample answer: I solved Problem 1 using U.S. traditional multiplication. Then I solved it again using partial-products multiplication. I got the same answer, so my answer makes sense.

For Problems 4–7, do the following:

- Write a number model with a letter for the unknown.
- Solve the problem. Use U.S. traditional multiplication for at least one problem. Show your work.
- Write the answer. **Sample number models given.**

4 Paula has 7 decks of cards. Each deck of cards has 52 cards in it. How many cards does she have in all?

Number model: __$52 * 7 = c$__

Answer: __364__ cards

5 A bush is 21 inches tall. A tree is 5 times as tall as the bush. How tall is the tree?

Number model: __$21 * 5 = x$__

Answer: __105__ inches

6 A fence has 45 sections. Each section is 6 meters long. How long is the fence?

Number model: __$45 * 6 = f$__

Answer: __270__ meters

7 An apartment building has 9 apartments on each floor. There are 43 floors. How many apartments are in the building?

Number model: __$43 * 9 = a$__

Answer: __387__ apartments

5.NBT.5, SMP1

Math Boxes

1 Solve.

a. $(3 \times 4) + (18 \div 3) =$ ___18___

b. $[7 \times (2 + 3)] - 20 =$ ___15___

c. $2 \times \{(81 \div 9) \div (9 \div 3)\} =$ ___6___

SRB 42-43

2 Write a 5-digit number with
6 in the ones place,
3 in the thousands place,
1 in the hundreds place,
8 in the ten-thousands place,
and 0 in the tens place.

$8\ 3\ ,\ 1\ 0\ 6$

SRB 66-67

3 Which expressions show 3,248 in expanded form?

Fill in the circle next to <u>all</u> that apply.

(A) $32 \times 1,000 + 4 \times 10 + 8 \times 1$

(B) 3 [1,000s] + 2 [100s] + 4 [10s] + 8 [1s] ●

(C) $3 \times 1,000 + 2 \times 100 + 4 \times 10 + 8 \times 1$

(D) $3 \times 10^3 + 2 \times 10^2 + 4 \times 10^1 + 8 \times 10^0$ ●

SRB 70

4 Find the volume of the prism.
Use the formula $V = l \times w \times h$.

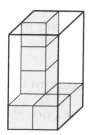

Volume = ___2___ × ___2___ × ___4___ = ___16___ units³

SRB 231-233

5 Write in exponential notation.

a. $10 \times 10 \times 10 \times 10$ ___10^4___

b. $10 \times 10 \times 10$ ___10^3___

c. $10 \times 10 \times 10 \times 10 \times 10 \times 10 \times 10 \times 10 \times 10 \times 10 \times 10$ ___10^{11}___

SRB 68

6 Asher used 5 apples to make an apple pie. To make a jar of applesauce he needed twice as many apples as he needed for the pie plus two more. Write an expression that models how many apples Asher needed for the applesauce.

Sample answer:
_____
$(5 \times 2) + 2$

SRB 42, 44

① 5.OA.1 ② 5.NBT.1 ③ 5.NBT.1 ④ 5.MD.5, 5.MD.5a, 5.MD.5b ⑤ 5.NBT.2 ⑥ 5.OA.2

49

Practicing with Powers of 10

SRB
68-69

In our place-value system the powers of 10 are grouped into sets of three, which are called periods. We have periods for ones, thousands, millions, billions, and so on. When we write large numbers in standard notation, we separate these periods with commas. Mathematical language includes prefixes for the periods and other important powers of 10.

Periods									
	Millions			Thousands			Ones		
Billions	Hundred-millions	Ten-millions	Millions	Hundred-thousands	Ten-thousands	Thousands	Hundreds	Tens	Ones
10^9	10^8	10^7	10^6	10^5	10^4	10^3	10^2	10^1	10^0

Use the place-value chart and the prefixes chart to complete the following statements and fill in the missing exponents.

Prefixes	
tera-	trillion (10^{12})
giga-	billion (10^9)
mega-	million (10^6)
kilo-	thousand (10^3)
hecto-	hundred (10^2)
deca-	ten (10^1)
uni-	one (10^0)

1. The distance from Chicago to New Orleans is about 10^3, or one __thousand__, miles.

2. A millionaire has at least $10^{\boxed{6}}$ dollars.

3. The Moon is about 240,000, or __24__ $* 10^{\boxed{4}}$, miles from Earth.

4. A computer with a 1-terabyte hard drive can store approximately $10^{\boxed{12}}$, or one __trillion__, bytes of information.

5. The Sun is about $89 * 10^7$, or __890,000,000__, miles from Saturn.

6. A 5-megapixel camera has a resolution of $5 * 10^{\boxed{6}}$, or 5 __million__ pixels.

7. What patterns do you notice in the following number sentences?

   $42 * 100 = 42 * 10^2 = 4,200$

   $42 * 1,000 = 42 * 10^3 = 42,000$

   $42 * 10,000 = 42 * 10^4 = 420,000$

   Sample answer: An extra zero is attached to the answer each time. The number of zeros is the same as the exponent.

Math Boxes

(1) Solve.

a. $3 * 100 =$ **300**

b. $6 * 1,000 =$ **6,000**

c. $8 * 10,000 =$ **80,000**

d. $3 *$ **100,000** $= 300,000$

e. $5 *$ **1,000,000** $= 5,000,000$

SRB
95-96

(2) Write each number in exponential notation.

a. $100 =$ 10^2

b. $10,000 =$ 10^4

c. $1,000,000 =$ 10^6

d. $100,000 =$ 10^5

e. $100,000,000 =$ 10^8

SRB
68

(3) Find the volume of the prism. Use the formula $V = B \times h$.

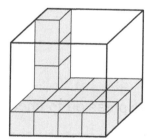

Area of the base = **12** units²

$V =$ **12** $\times$ **4** $=$ **48** units³

SRB
231-233

(4) Make an estimate and then solve.

Estimates vary.

a. _____   **b.** _____
   (estimate)   (estimate)

$$\begin{array}{r} 3\ \ 4 \\ *\quad 8 \\ \hline 2\ 7\ 2 \end{array} \qquad \begin{array}{r} 5\ \ 2 \\ *\quad 6 \\ \hline 3\ 1\ 2 \end{array}$$

SRB
83, 100,
102, 104

(5) **Writing/Reasoning** Explain how you knew how many zeros belonged in the products in Problems 1a–1c.

Sample answer: The product has the same number of zeros as the second number I multiplied because it is a power of 10.

SRB
27-29,
95-96

① 5.NBT.2 ② 5.NBT.2 ③ 5.MD.5, 5.MD.5a, 5.MD.5b
④ 5.NBT.5 ⑤ 5.NBT.2, SMP8

51

1 A dairy worker has 12 gallons of milk. She wants to pour it into 1-quart containers.

a. How many quarts are in 1 gallon?

__4__ quarts

b. How many quarts are in 12 gallons?

__48__ quarts

c. How many quarts of milk does the dairy worker have?

__48__ quarts

d. Write an expression to model the number story.

12 * 4

2 A seamstress sewed two pieces of fabric together. One piece was 3 feet long. The other was 8 inches long.

SRB
44, 215-216, 328

a. What is the length of the 3-foot piece of fabric in inches?

__36__ inches

b. What is the total length of the new piece of fabric?

__44__ inches

c. Write an expression to model both steps of the number story.

(3 * 12) + 8

For Problems 3 and 4:

- Solve the problem.
- Write an expression to model the problem. Evaluate the expression to check your answer.

3 Two fifth-grade students had a running race. It took one student 2 minutes to run from one end of the playground to the other. It took the other student 98 seconds. How much faster was the second student's time?

Try This

4 A restaurant chef has 5 pounds of steak. He wants to cut it into 8-ounce portions to serve to customers. How many 8-ounce portions can he make?

Answer: __22__ seconds

(2 * 60) − 98

(number model)

Answer: __10__ portions

(5 * 16) / 8

(number model)

52   5.OA.1, 5.OA.2, 5.MD.1, SMP1, SMP4

Math Boxes

1 Complete.

a. $4 \times 3 =$ _____12_____

b. $3 \times 10^3 =$ _____3,000_____

c. $4 \times$ _____3,000_____ $= 12,000$

d. $4 \times 3 \times$ _____10^3,_____ $= 12,000$

or 1,000

SRB
95-98

2 The figure below is a mathematical model of a blanket fort Sue built in her bedroom. Use the model to estimate the volume of the fort.

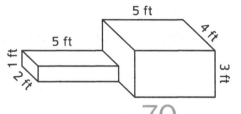

Volume: About _____70_____ cubic feet

SRB
233-234

3 Make an estimate and solve.

$492 * 4 = ?$

_____Answers vary._____
(estimate)

```
      4   9   2
  *           4
  _____
  1, 9 6 8
```

SRB
83, 100,
102, 104

4 Insert grouping symbols to make the number sentences true.

a. $5 * (4 - 2) = 10$

b. $45 / (9 + 6) = 3$

c. $2 * \{(4 \div 2) + 3\} = 10$

SRB
42-43

5 Which 6-digit numbers have
2 in the ones place,
7 in the thousands place, and
9 in the hundred-thousands place?

☐ 942,147   ☐ 749,124

☑ 947,142   ☐ 497,142

☑ 927,442

SRB
66-67

6 Fill in the missing digits.

```
      2
    7   6
  *       4
  _____
  3 0 4
```

SRB
102

① 5.NBT.2 ② 5.MD.5, 5.MD.5b, 5.MD.5c ③ 5.NBT.5
④ 5.OA.1 ⑤ 5.NBT.1 ⑥ 5.NBT.5

53

For Problems 1–3:

• Make an estimate. Write a number model to show how you estimated.
• Solve using U.S. traditional multiplication. Show your work.
• Use your estimate to check whether your answer makes sense.

Sample estimates given.

SRB 83, 103

Example: 76 * 24 = ?

Estimate: **80 * 20 = 1,600**

```
      1
      2
      7  6
 *    2  4
 ───────────
      3  0  4
 +  1, 5  2  0
 ───────────
    1, 8  2  4
```

1 31 * 43 = ?

Estimate: 30 * 40 = 1,200

```
      3  1
 *    4  3
 ───────────
      9  3
 +  1, 2  4  0
 ───────────
    1, 3  3  3
```

2 26 * 16 = ?

Estimate: 25 * 20 = 500

```
      3
      2  6
 *    1  6
 ───────────
    1  5  6
 +  2  6  0
 ───────────
    4  1  6
```

3 87 * 46 = ?

Estimate: 90 * 50 = 4,500

```
      2
      4
      8  7
 *    4  6
 ───────────
      5  2  2
 +  3, 4  8  0
 ───────────
    4, 0  0  2
```

4 Explain how your estimate helps you check whether your answer in Problem 1 makes sense. Sample answer: My estimate was 1,200 and my answer was 1,333. Those numbers are pretty close together, so my answer makes sense.

Try This

Sample explanation:

5 Complete the area model. Explain how it relates to your work for Problem 3.

The area of the right side of the rectangle is my first partial product, or 522. The area of the left side of the rectangle is my second partial product, or 3,480.

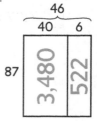

Math Boxes

1 Complete. Use exponential notation for Parts d and e.

a. $8 \times 10^2 =$ ___800___

b. $3 \times 10^3 =$ ___3,000___

c. $5 \times 10^4 =$ ___50,000___

d. $2 \times$ ___10^6___ $= 2,000,000$

e. $7 \times$ ___10^5___ $= 700,000$

SRB 95-96

2 Complete the table.

SRB 68

Standard Notation	Exponential Notation
10,000	10^4
100,000	10^5
10,000,000	10^7
1,000,000	10^6

3 Find the volume of the prism. Use the formula $V = B \times h$.

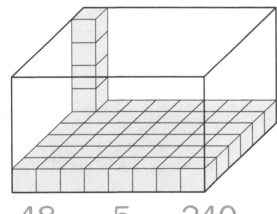

$V =$ ___48___ $\times$ ___5___ $=$ ___240___ units^3

SRB 231-233

4 Make an estimate and solve.

Estimates vary.

a. _____   b. _____
      (estimate)   (estimate)

```
    6  4            8  9
 *     5         *     3
 ----------      ----------
 3 2 0           2 6 7
```

SRB 83, 100, 102, 104

5 **Writing/Reasoning** Explain how you know the formula $V = B \times h$ tells you how many unit cubes could be packed into the prism in Problem 3. Sample answer:

The area of the base tells you how many cubes fit in one layer of the prism. The height tells you how many layers there are. When you multiply the cubes in one layer by the number of layers, you find out how many cubes could be packed in the whole prism.

SRB 233

① 5.NBT.2 ② 5.NBT.2 ③ 5.MD.5, 5.MD.5a, 5.MD.5b
④ 5.NBT.5 ⑤ 5.MD.5, 5.MD.5a, SMP2, SMP8

Choosing Multiplication Strategies

Solve Problem 1 using U.S. traditional multiplication. Solve Problems 2–6 using any strategy. Show your work. Use your estimates to check whether your answers make sense.

SRB 83, 100-104

① 627 * 34 = ?

Estimate: 600 * 30 = 18,000

② 148 * 8 = ?

Estimate: 148 * 10 = 1,480

Sample estimates given.

627 * 34 = __21,318__

148 * 8 = __1,184__

③ 72 * 110 = ?

Estimate: 70 * 100 = 7,000

④ 436 * 65 = ?

Estimate: 400 * 70 = 28,000

72 * 110 = __7,920__

436 * 65 = __28,340__

⑤ A clerk ordered 72 boxes of paper clips. There are 250 paper clips in each box. How many paper clips are there in all?

Estimate: 70 * 300 = 21,000

⑥ A photo of a building is 18 centimeters tall. The real building is 892 times as tall. How tall is the building?

Estimate: 20 * 900 = 18,000

Answer: __18,000__ paper clips

Answer: __16,056__ centimeters

⑦ What strategy did you use to solve Problem 3? Explain why you chose that strategy.

Sample answer: I used partial-products multiplication because I knew I could find the partial products mentally.

Math Boxes

1 Complete.

a. $7 \times 2 =$ ___14___

b. $2 \times 10^2 =$ ___200___

c. $7 \times 10^2 =$ ___700___

d. $(7 \times 10^2) \times (2 \times 10^2) =$
   ___140,000___

e. $700 \times 200 =$ ___140,000___

SRB
95-98

2 The figure below is a mathematical model of Remy's toy train car. Use the model to estimate the volume of the train car.

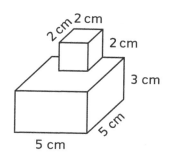

Volume: About ___83___ cm³

SRB
233-234

3 Make an estimate and then solve.

a. ___Estimates vary.___   b. _____
    (estimate)   (estimate)

```
   2  8  7          4  1
*        9       * 1  7
2, 5 8 3          6 9 7
```

SRB
83, 100,
102, 104

4 Insert grouping symbols to make the number sentences true.

a. $36 / (6 - 5) = 36$

b. $25 * (8 - 3) = 125$

c. $(2 + 36) \div (12 - 10) = 19$

d. $2 + [36 \div (12 - 6)] = 8$

SRB
42-43

5 Write a 7-digit number with
5 in the hundred-thousands place,
2 in the tens place,
4 in the millions place,
6 in the ten-thousands place, and
0s in the other places.

___4,560,020___

SRB
66-67

6 Fill in the missing digits.

```
        1
      6  2
   ×       8
   4  9  6
```

SRB
102

① 5.NBT.2 ② 5.MD.5, 5.MD.5b, 5.MD.5c ③ 5.NBT.5
④ 5.OA.1 ⑤ 5.NBT.1 ⑥ 5.NBT.5

Invitations

Zoey is mailing invitations for a fifth-grade party. It takes her about 30 seconds to address 1 envelope.

1 About how many seconds would it take Zoey to address 10 envelopes? Show your work.

About __300__ seconds

Sample answers:
- 10 envelopes * 30 seconds per envelope = 300 seconds
- 10 * 30 = 300 seconds

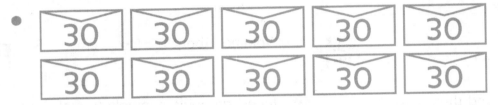

2 About how many seconds would it take Zoey to address 100 envelopes? Show your work.

About __3,000__ seconds

Sample answers:
- In Problem 1 I found that it takes 300 seconds to address 10 envelopes. There are 10 times as many envelopes to address in Problem 2 as in Problem 1.
  10 * 300 = 3,000 seconds
- 30 seconds per envelope * 100 envelopes = 3,000 seconds

5.NBT.2, 5.NBT.5

Math Boxes

1 It took 32 minutes for Tara to walk to the store, 56 minutes to do her shopping, and 32 minutes to walk home. How many hours was Tara gone? Sample

answer: (32 + 56 + 32) / 60

(number model)

Answer: ___2___ hours

SRB
44, 216

2 Make an estimate and solve.

a.
```
    3  1  2
  ×    2  3
```
Estimates vary.

(estimate)

7, 1 7 6

b.
```
    4  9  6
  ×    3  2
```

(estimate)

1 5, 8 7 2

SRB
83, 100, 104

3 Which expression shows 5,892 in expanded form?
Fill in the circle next to the best answer.

- **A.** $(5 \times 10^3) + (8 \times 10^2) + (9 \times 10^1) + (2 \times 10^0)$

- **B.** $(5 \times 10^4) + (8 \times 10^3) + (9 \times 10^2) + (2 \times 10^1)$

- **C.** $(5 \times 10^1) + (8 \times 10^2) + (9 \times 10^3) + (2 \times 10^4)$

SRB
70

4 Find the volume of the prism.
Use the formula $V = l \times w \times h$.

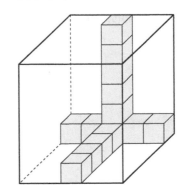

$V = \underline{5} \times \underline{5} \times \underline{6} = \underline{150}$ units³

SRB
231-233

5 **Writing/Reasoning** What method did you use to multiply in Problem 2a? Why did you choose that method?

Sample answer: I used partial products to make it easier to keep track of the parts I was multiplying.

SRB
100-104

① 5.MD.1 ② 5.NBT.5 ③ 5.NBT.1, 5.NBT.2
④ 5.MD.5, 5.MD.5a, 5.MD.5b ⑤ 5.NBT.5, SMP6

Using Multiples to Rename Dividends

Draw 2 cards to create a dividend. Draw one more card to create a divisor. Use multiples of the divisor to make equivalent names for the dividend. Then solve the division problem. Summarize your solution.

SRB
72, 107

Example: Rosie drew 7, 4, and 3 and made her dividend 74 and her divisor 3. She listed multiples of 3 and completed the following name-collection box.

74
30 + 30 + 14
60 + 14
60 + 12 + 2

Multiples of 3: 3, 6, 9, 12, 15, 18, 21, 24, 27, 30...

$$60 + 12 + 2$$

$$60 / 3 = 20 \qquad 12 / 3 = 4$$

$$74 / 3 \rightarrow 24 \text{ R2}$$

Answers vary.

---

① Dividend: _____ Divisor: _____

Multiples of divisor: _____

_____

Summary: _____ / _____ → _____

---

② Dividend: _____ Divisor: _____

Multiples of divisor: _____

_____

Summary: _____ / _____ → _____

---

③ Dividend: _____ Divisor: _____

Multiples of divisor: _____

_____

Summary: _____ / _____ → _____

---

④ Dividend: _____ Divisor: _____

Multiples of divisor: _____

_____

Summary: _____ / _____ → _____

Practicing Unit Conversions

For Problems 1 and 2, complete the tables to show unit conversions.

SRB
44, 215–
217, 328

1

Hours	Minutes
1	60
2	120
5	300
	Answers vary.

2

Yards	Inches
1	36
3	108
10	360
	Answers vary.

For Problems 3–6: Sample number models given.

- Solve the problem.
- Write an expression to model the problem.
- Evaluate the expression to check your answer.

3 An adult African elephant can weigh up to 7 tons. At birth an African elephant weighs about 200 pounds. How many more pounds does an adult African elephant weigh than a newborn?

Answer: __13,800__ pounds

$(7 * 2,000) - 200$

(number model)

4 A landscaper ordered 6 cubic yards of soil. So far she has used 90 cubic feet of soil. How many cubic feet of soil are left?

Hint: How many cubic feet are in 1 cubic yard?

Answer: __72__ cubic feet

$(6 * 27) - 90$

(number model)

5 A rectangular room is 4 yards long and 5 yards wide. Lucas is covering the floor with tiles that are 1 square foot. How many tiles will he need?

Hint: Start by finding the area of the room in square yards.

Answer: __180__ tiles

$(4 * 5) * 9$

(number model)

Try This

6 A football coach mixed 4 gallons of sports drink for his team. A serving of sports drink is 1 cup. How many servings of sports drink did the coach mix?

Answer: __64__ servings

$4 * 4 * 2 * 2$

(number model)

5.OA.1, 5.OA.2, 5.MD.1, SMP1, SMP4

Math Boxes: Preview for Unit 3

① Which model shows $\frac{5}{6}$ shaded?

Choose the best answer.

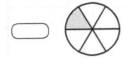
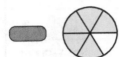

SRB
153, 155

② Yasmin had 3 bananas. She shared them equally among the 4 people in her family. Write an expression that shows how much banana each person got.

$$3 \div 4$$

SRB
38, 44,
163-164

③ Marcus spent $\frac{1}{2}$ of his allowance on trading cards and $\frac{1}{4}$ of his allowance on snacks. Did he spend more on trading cards or snacks?

trading cards

SRB
174-175

④ Write two fractions equivalent to $\frac{1}{2}$.

$$\frac{4}{8} \qquad \frac{50}{100}$$

Write two fractions equivalent to $\frac{1}{4}$.

$$\frac{3}{12} \qquad \frac{25}{100}$$

Sample answers given.

SRB
165-166,
168, 170

⑤ Solve.

a. $\dfrac{1}{8} + \dfrac{1}{8} + \dfrac{1}{8} = \dfrac{3}{8}$

b. $\dfrac{1}{4} + \dfrac{1}{4} + \dfrac{1}{4} = \dfrac{3}{4}$

c. $\dfrac{1}{5} + \dfrac{3}{5} = \dfrac{4}{5}$

d. $\dfrac{2}{9} + \dfrac{5}{9} = \dfrac{7}{9}$

SRB
186

⑥ Brielle is buying yarn to knit a scarf. She needs to know the area of the scarf she will knit to choose the right package of yarn. What is the area of a scarf that is 4 feet long and $\frac{1}{2}$ foot wide?

4 ft

 $\frac{1}{2}$ ft

Area = 2 square feet

SRB
186, 225

① 5.NF.3 ② 5.OA.2, 5.NF.3 ③ 5.NF.2 ④ 5.NF.1 ⑤ 5.NF.1

62 ⑥ 5.NF.4, 5.NF.4b

Partial-Quotients Division

For Problems 1–4, make an estimate. Then use partial-quotients division to solve. Show your work on the computation grid. Sample estimates given.

SRB 84, 108-112

1 234 / 11 → ?

Estimate: _200 / 10 = 20_

Answer: _21 R3_

2 825 / 15 → ?

Estimate: _800 / 20 = 40_

Answer: _55 R0_

3 3,518 / 30 → ?

Estimate: _3,600 / 30 = 120_

Answer: _117 R8_

4 6,048 / 54 → ?

Estimate: _6,000 / 60 = 100_

Answer: _112 R0_

Try This

5 Complete the area model on the right to show your solution for Problem 2.

Hint: Think of Problem 2 as: If the area of a room is 825 square feet and the length of the room is 15 feet, how wide is the room?

Sample answer:

Area (Dividend): _825_
Length (Divisor): _15_

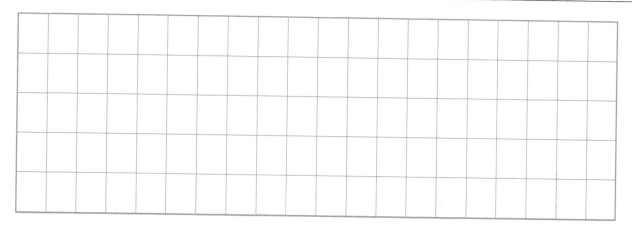

150	10
150	10
150	10
150	10
150	10
75	5

Width (Quotient): 55

5.NBT.6, SMP2

63

1 Solve.

a. $45 / 9 =$ __5__

b. $450 / 9 =$ __50__

c. $4,500 / 9 =$ __500__

d. $32 / 8 =$ __4__

e. $320 / 8 =$ __40__

f. $3,200 / 8 =$ __400__

SRB 106

2 Find the volume of the prism.
Use the formula $V = B \times h$.

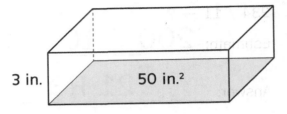

3 in.   50 in.²

$V =$ __50__ $\times$ __3__ $=$ __150__ in.³

SRB 233

3 True or false?

In the number 23,916:

a. the digit 3 is worth 3,000.

　● true　　○ false

b. the digit 9 is worth 90.

　○ true　　● false

c. the digit 2 is worth 20,000.

　● true　　○ false

d. the digit 1 is worth 100.

　○ true　　● false

SRB 66-67

4 Fill in the missing digits.

a.
```
        4   1
    2   8   2
×           6
_____
1,  6   9   2
```

b.
```
    4   3
3   8   6
×       5
_____
1,  9   3   0
```

SRB 102

5 **Writing/Reasoning** Explain how you solved Problem 1e.

Sample answer: I know 32 divided by 8 is 4. Since 320 is 10 times more than 32, I multiplied 4 by 10 to get 40.

SRB 106

① 5.NBT.6 ② 5.MD.5, 5.MD.5b ③ 5.NBT.1 ④ 5.NBT.5
⑤ 5.NBT.6, SMP6, SMP7

Partial Quotients with Multiples

For Problems 1–4, make an estimate. Then use partial-quotients division to solve. Show your work. You can make lists of multiples on *Math Masters,* page TA10 to help you.

Sample estimates given.

① 1,647 / 28 → ?

Estimate: __1,500 / 30 = 50__

Answer: __58 R23__

② 4,319 / 42 → ?

Estimate: __4,200 / 42 = 100__

Answer: __102 R35__

③ 2,628 / 36 → ?

Estimate: __2,800 / 40 = 70__

Answer: __73 R0__

④ 9,236 / 41 → ?

Estimate: __8,000 / 40 = 200__

Answer: __225 R11__

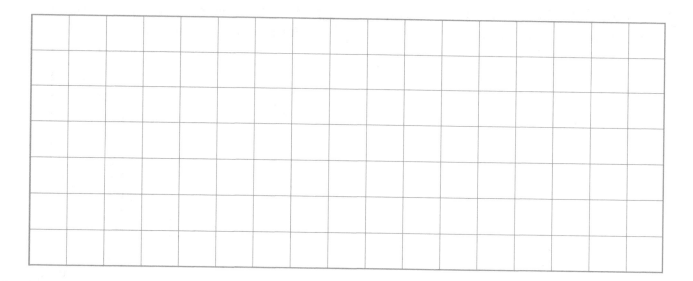

Try This

⑤ Paul drew the area model at the right for his solution to Problem 1. What partial quotients did he use to solve the problem?

__50, 5, 2, and 1__

Area (Dividend): 1,647

Length (Divisor): 28

1,400	50
140	5
56	2
28	1
23	

Width (Quotient): about 58

5.NBT.6

65

Math Boxes

1 Yao bought 3 feet of blue ribbon and 24 inches of red ribbon. The ribbon costs $2.00 per foot. What is the total cost of the ribbon? Sample answer:

$$[(24 / 12) + 3] * 2$$

(number model)

Answer: _**10**_ dollars

SRB
44, 216

2 Fill in the circle next to the best estimate for the problem below. Then solve.

Estimate:

$$\begin{array}{r} 2\ 1\ 7 \\ \times\ 1\ 9\ 8 \\ \hline 42{,}966 \end{array}$$

Ⓐ 40,000

Ⓑ 4,000

Ⓒ 20,000

Ⓓ 400,000

SRB
83,
100-104

3 Write 72,658 in expanded form using exponents to write powers of 10.

$$(7 \times 10^4) + (2 \times 10^3) +$$

$$(6 \times 10^2) + (5 \times 10^1) +$$

$$(8 \times 10^0)$$

SRB
70

4 Find the volume of the prism. Use the formula $V = l \times w \times h$.

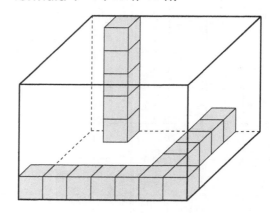

$V = \underline{7} \times \underline{6} \times \underline{5} = \underline{210}$ units³

SRB
231-233

5 **Writing/Reasoning** Explain how you solved Problem 1.

Sample answer: I changed 24 inches to 2 feet. Then I added 3 feet and 2 feet and got 5 feet. I multiplied 5 feet by $2 per foot. I got $10.00.

SRB
44, 216

① 5.MD.1 ② 5.NBT.5 ③ 5.NBT.1, 5.NBT.2
④ 5.MD.5, 5.MD.5a, 5.MD.5b ⑤ 5.MD.1, SMP1, SMP6

Math Boxes

Math Boxes

① Solve.

a. 42 / 6 = **7**

b. 420 / 6 = **70**

c. 4,200 / 60 = **70**

d. 81 / 9 = **9**

e. 81,000 / 90 = **900**

SRB
106

② Find the volume of the prism.
Use the formula $V = B \times h$.

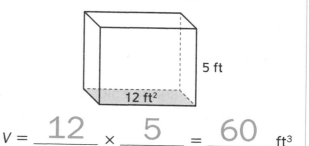

5 ft

12 ft²

$V =$ **12** × **5** = **60** ft³

SRB
233

③ Write the value of the boldface digit in each number.

a. 3**9**0 **90**

b. **8**,092 **8,000**

c. 3**5**,047 **5,000**

d. 2**3**2,591 **30,000**

e. **4**97,214 **400,000**

SRB
66-67

④ Fill in the missing digits.

a.
```
        2   1
    4   5   3
×           4
1,  8   1   2
```

b.
```
        2
3   2   7
×           3
9   8   1
```

SRB
102

⑤ Writing/Reasoning In Problem 3, how would the value of the boldface digits change if they moved one place to the right?

Answers vary: The digits would be worth $\frac{1}{10}$ of their value if they moved one place to the right.

SRB
66-67

① 5.NBT.6 ② 5.MD.5, 5.MD.5b ③ 5.NBT.1 ④ 5.NBT.5
⑤ 5.NBT.1, SMP7

Interpreting Remainders

SRB
12-14,
113

For each problem:

- Create a mathematical model.
- Solve the problem. Show your work.
- Tell what the remainder represents.
- Decide what to do with the remainder. Explain what you did.

Sample models shown for Problems 1–3.

---

① Basketballs are on sale for $12, including tax. How many basketballs can the gym teacher buy with $40?

Mathematical model:

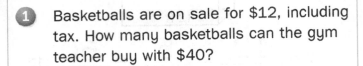

$4 left over

Quotient: ___3___ Remainder: ___4___

What does the remainder represent?

the $4 left over
_____

_____

Answer: The gym teacher can buy

___3___ basketballs.

Circle what you did with the remainder.

(Ignored it)

Rounded the quotient up

Why?

The $4 left over isn't
enough to buy another
basketball.

---

② You are organizing a trip to a museum for 110 students, teachers, and parents. If each bus can seat 25 people, how many buses do you need?

Mathematical model:

25 on each bus

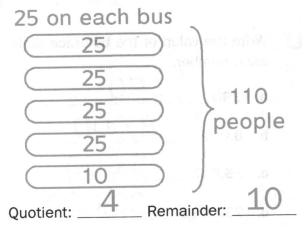

Quotient: ___4___ Remainder: ___10___

What does the remainder represent?

the number of people remaining
after 4 buses are filled

Answer: I need ___5___ buses.

Circle what you did with the remainder.

Ignored it

(Rounded the quotient up)

Why?

4 buses hold 100 people.
One more bus was needed
for the extra 10 people.

Interpreting Remainders
(continued)

Lesson 2-13

DATE   TIME

SRB
12-14,
113

3 Mrs. Maxwell has 60 pens to pass out to her class. There are 27 students in her class. How many pens will each student get if everyone is given a fair share?

Mathematical model:

$$60 \div 27 = p$$

Quotient: ___2___ Remainder: ___6___

What does the remainder represent?
the 6 pens left over
_____

Answer: Each student will get
___2___ pens.

Circle what you did with the remainder.

(Ignored it)

Rounded the quotient up

Why?
The 6 pens left are not
enough to give each
student one more pen.

4 Choose one of your mathematical models. Explain how it helped you solve the problem.

Answers vary.
_____
_____
_____
_____
_____
_____

1 Match the letters to the fractions they represent on the number line.

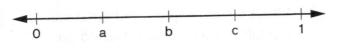

$\frac{1}{2}$ = _____ b

$\frac{1}{4}$ = _____ a

$\frac{3}{4}$ = _____ c

SRB
153, 158, 161

2 Ben has 3 cans of food to feed his cat for 5 days. Write an expression that shows how much of a can Ben should feed his cat each day.

$3 \div 5$, or $\frac{3}{5}$

SRB
38, 44, 163-164

3 Katie and Jonah were sharing a bag of pretzels. Katie had $\frac{1}{2}$ of the pretzels. Jonah had $\frac{1}{3}$ of the pretzels. Who had more pretzels?

Katie

SRB
174-175

4 Write two fractions equivalent to $\frac{1}{3}$.

$\frac{3}{9}$ $\frac{25}{75}$

Write two fractions equivalent to $\frac{2}{3}$.

$\frac{8}{12}$ $\frac{50}{75}$

 Sample answers given.

SRB
165-166, 168, 170

5 Solve.

a. $\frac{1}{7} + \frac{1}{7} + \frac{1}{7} =$ $\frac{3}{7}$

b. $\frac{1}{12} +$ $\frac{5}{12}$ $= \frac{6}{12}$

c. $\frac{3}{6} + \frac{1}{6} +$ $\frac{1}{6}$ $= \frac{5}{6}$

d. $\frac{3}{9}$ $+ \frac{2}{9} = \frac{5}{9}$

e. $\frac{1}{4} + \frac{1}{4} + \frac{1}{4} =$ $\frac{3}{4}$

SRB
186

6 Ricardo wants to cover a shelf with shelf liner. The shelf is 4 feet wide and $\frac{2}{3}$ feet deep. What is the area of the shelf?

Area = $2\frac{2}{3}$ square feet

SRB
12-14, 186, 225

① 5.NF.3 ② 5.OA.2, 5.NF.3 ③ 5.NF.2 ④ 5.NF.1
⑤ 5.NF.1 ⑥ 5.NF.4, 5.NF.4b

Solving Fair Share Number Stories

Use fraction circle pieces or a drawing to model each number story. Then solve.

SRB
163-164

1 Mary and her two friends were working on a science project. They shared 1 pizza equally as a snack. How much pizza did each person get?

Models: Sample models shown.

Solution: _____ $\frac{1}{3}$ pizza _____

2 Jose is taking care of a neighbor's cat. The neighbor will be gone for 5 days and left 3 cans of cat food. The cat is supposed to eat the same amount each day. How much food should Jose give the cat each day?

Solution: _____ $\frac{3}{5}$ can _____

3 A school received a shipment of 4 boxes of paper. The school wants to split the paper equally among its 3 printers. How much paper should go to each printer?

Solution: $\frac{4}{3}$, or $1\frac{1}{3}$, boxes of paper

4 Adrian brought 2 loaves of olive bread to school for a class celebration. There were 12 people who wanted to try the bread. They decided to split the loaves evenly. How much bread did each person receive?

Solution: _____ $\frac{2}{12}$, or $\frac{1}{6}$, loaf _____

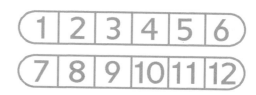

5.NF.3, SMP4

71

Multiplication and Division

For Problems 1–3, make an estimate. Then solve using U.S. traditional multiplication. Use your estimates to check whether your answers make sense. Sample estimates given.

SRB
83, 103,
108-112

1 30 * 60 = 1,800
(estimate)

```
     4
     5
    2 8
  * 5 7
  1 9 6
+ 1 4 0 0
  1, 5 9 6
```

2 600 * 80 = 48,000
(estimate)

```
      3 2
     6 4 3
   *   8 1
     6 4 3
 + 5 1 4 4 0
   5 2,0 8 3
```

3 700 * 150 = 105,0
(estimate)

```
        2
        3
      7 0 6
    * 1 4 5
    3 5 3 0
  2 8 2 4 0
+ 7 0 6 0 0
1 0 2,3 7 0
```

For Problems 4 and 5, make an estimate. Then solve using partial-quotients division. Use your estimates to check whether your answers make sense. Sample estimates given.

4 4,500 ÷ 30 = 150
(estimate)

```
      130   R8
32)4, 1 6 8
 -3,200 | 100
    968
  -640  | 20
    328
  -320  | 10
      8 | 130
```

Sample partial quotients are given.

Answer: ___130___ R___8___

5 7,000 ÷ 50 = 140
(estimate)

```
      128   R43
56)7, 2 1 1
 -5,600 | 100
  1,611
 -1,120 | 20
    491
  -280  | 5
    211
  -112  | 2
     99
   -56  | 1
     43 | 128
```

Answer: ___128___ R___43___

6 Complete the area model to represent your solution to Problem 4.

Area (Dividend): 4,168

Length (Divisor): 32

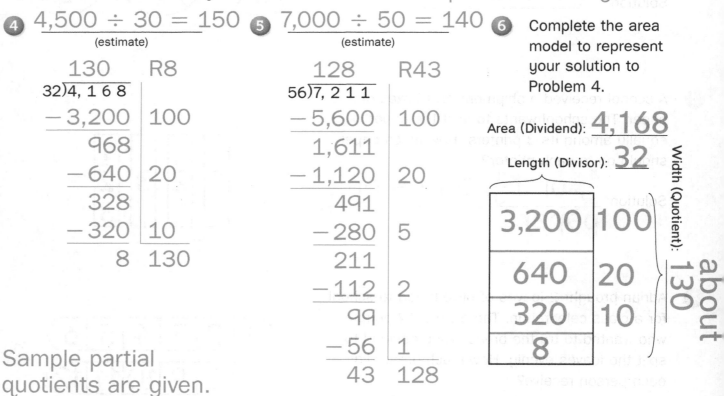

Width (Quotient): 130

about 130

Sample model given.

1 Insert grouping symbols to make true number sentences.

a. $(19 + 41) * 3 = 180$

b. $5 = (16 / 2) + 2 - 5$

c. $24 \div (8 + 4) * 3 = 6$

d. $(24 \div 8) + (4 * 3) = 15$

e. $1 = \{16 / (2 + 2)\} - 3$

SRB 42-43

2 Complete the following equivalents.

a. 1 pint = ___2___ cups

b. 6 quarts = ___12___ pints

c. 1 quart = ___4___ cups

d. 3 gallons = ___12___ quarts

e. 1 gallon = ___16___ cups

SRB 215-217, 328

3 Write in standard notation or expanded form. Sample answer:

a. $82{,}913 = $ ___80,000 + 2,000 + 900 + 10 + 3___

b. $2 \times 100{,}000 + 6 \times 10{,}000 + 1 \times 1{,}000 + 9 \times 100 + 4 \times 10 + 5 \times 1 = $ ___261,945___

c. $5 \times 10^3 + 2 \times 10^2 + 0 \times 10^1 + 7 \times 10^0 = $ ___5,207___

SRB 70

4 What is the volume of the prism?

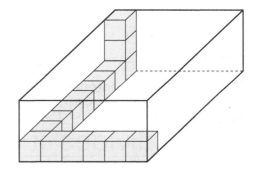

Volume = ___162___ cubic units

SRB 231-233

5 Solve mentally by breaking the dividend into smaller parts that are easier to divide. Write the equivalent name you used.

$96 \div 3 \rightarrow$ ___32___

Equivalent name for 96:

Sample answer:

___90 + 6___

SRB 107

6 Solve. Use an estimate to check whether your answer makes sense.

$$\begin{array}{r} 7\,2\,8 \\ \times\ 1\,2 \\ \hline 8{,}736 \end{array}$$

___Answers vary.___
(estimate)

SRB 83, 100-104

① 5.OA.1 ② 5.MD.1 ③ 5.NBT.1 ④ 5.MD.3, 5.MD.3b, 5.MD.4
⑤ 5.NBT.6 ⑥ 5.NBT.5

Writing Division Number Stories

Record the fractions you are assigned. For each fraction, write a division number sentence with the fraction as the quotient. Then write a number story to match the number sentence. SRB 163-164

Sample answers given.

1. Fraction: $\frac{7}{8}$

   Division number sentence: $7 \div 8 = \frac{7}{8}$

   Number story: There were 7 granola bars in a box. If 8 people split the granola bars equally, how much of a granola bar will each person get?

2. Fraction: $\frac{5}{4}$

   Division number sentence: $5 \div 4 = \frac{5}{4}$

   Number story: Heather has 5 yards of fabric. If she divides it into 4 equal pieces, how many yards will each piece be?

3. Fraction: $\frac{5}{6}$

   Division number sentence: $5 \div 6 = \frac{5}{6}$

   Number story: Coach Ray brought 5 liters of juice to practice. If he pours the same amount of juice into 6 glasses, how many liters of juice will there be in each glass?

Solve each number story. You can use fraction circle pieces or drawings to help. Write a number model to show how you solved each problem.

SRB
163-164

1 Davita brought 6 granola bars for herself and the 7 other girls in her camp group for a snack. If they share them equally, what fraction of a granola bar will each girl get?

Solution: $\frac{6}{8}$, or $\frac{3}{4}$ granola bar

Number model: $6 \div 8 = \frac{6}{8}$, or $6 \div 8 = \frac{3}{4}$

2 Lucas is making 12 jumbo muffins to sell at his class bake sale. He has 2 bowls full of batter. What fraction of a bowl of batter should Lucas put in each muffin cup?

Solution: $\frac{2}{12}$, or $\frac{1}{6}$ bowl

Number model: $2 \div 12 = \frac{2}{12}$, or $2 \div 12 = \frac{1}{6}$

3 Ms. Cox is combining bottles of hand sanitizer. She has 11 small bottles of sanitizer she wants to divide equally among 3 large containers. How many small bottles should she empty into each large container?

Solution: $\frac{11}{3}$, or $3\frac{2}{3}$ small bottles

Number model: $11 \div 3 = \frac{11}{3}$, or $11 \div 3 = 3\frac{2}{3}$

4 Write a division number story with an answer of $\frac{12}{8}$.

Sample answer: Michelle brought 12 apples to share at her dance practice. There were 7 other girls besides Michelle. The girls decided to split the apples equally. How many apples will each girl get?

Number model: $12 \div 8 = \frac{12}{8}$, or $12 \div 8 = 1\frac{1}{2}$

Math Boxes

① Solve.

 a. $4 \times 100 =$ _**400**_

 b. $4 \times 10^2 =$ _**400**_

 c. $6 \times 10^3 =$ _**6,000**_

 d. $6 \times 1,000 =$ _**6,000**_

> **SRB**
> 95-96

② Solve.

$25\overline{)578}$

$578 \div 25 \rightarrow$ _**23 R3**_

> **SRB**
> 108-110

③ Find the area of a table top that is $3\frac{1}{3}$ feet by 2 feet.

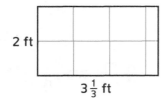

2 ft

$3\frac{1}{3}$ ft

Area = _**$6\frac{2}{3}$**_ square feet

Sample answer:
(number model)

$3\frac{1}{3} + 3\frac{1}{3} = 6\frac{2}{3}$

> **SRB**
> 224-225

④ Kayin buys 6 envelopes for 35 cents each and 6 stamps for 48 cents each. Which expression models how much money Kayin spends?

Fill in the circle next to the best answer.

◯ **A.** $(6 + 6) * (35 + 48)$

◯ **B.** $(6 * 6) + (35 + 48)$

⬤ **C.** $(6 * 35) + (6 * 48)$

> **SRB**
> 42, 44

⑤ **Writing/Reasoning** Describe a pattern you noticed in Problem 1.

Sample answer: The number in the exponent is the same as the number of zeros in the product.

> **SRB**
> 95-96

Division Number Stories with Remainders

For each number story:

- Write a number model with a letter for the unknown.
- Solve. Show your work in the space provided. You may draw a picture to help.
- Decide what to do with the remainder. Explain what you did and why.

SRB
44,
113-114

1 Rebecca and her two sisters made pancakes for breakfast. They made 16 pancakes for 5 people. They want to make sure each person gets an equal serving. How many pancakes will each person get? *Sample answer:*

Number model: $16 \div 5 = p$

Sample work:

$3 * 5 = 15,$
so $16 \div 5 = 3$ R1

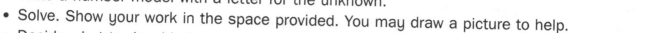

Quotient: __3__ Remainder: __1__

Answer: Each person will get __$3\frac{1}{5}$__ pancakes.

Circle what you did with the remainder.

　Ignored it

　(Reported it as a fraction)

　Rounded the quotient up

Why? *Sample answer:*
Pancakes can be cut into smaller pieces.

2 Louis's soccer team is taking a bus to a tournament. They have 32 reusable water bottles. Their water carriers hold 6 bottles each. How many carriers will Louis's team need to bring all of their water bottles on the bus? *Sample answer:*

Number model: $32 \div 6 = c$

Sample work:

| 6 bottles | 6 bottles | 6 bottles |
| 6 bottles | 6 bottles | |

Quotient: __5__ Remainder: __2__

Answer: Louis's team needs __6__ carriers.

Circle what you did with the remainder.

　Ignored it

　Reported it as a fraction

　(Rounded the quotient up)

Why? *Sample answer: They*
needed another carrier for the two bottles left over.

5.NBT.6, 5.NF.3, SMP1, SMP6

3 Mariana saved $80 from her babysitting job. She wants to buy some shirts and pants that are on sale at her favorite store for $17 each. How many items of clothing can she buy?

Sample answer:

Number model: $\underline{80 \div 17 = i}$

Sample work:

$$17\overline{)80}$$
$$\underline{-\ 34} \quad 2$$
$$46$$
$$\underline{-\ 34} \quad 2$$
$$12 \quad 4$$

Quotient: ___4___ Remainder: ___12___

Answer: Mariana can buy

___4___ items.

Circle what you did with the remainder.

(Ignored it)

Reported it as a fraction

Rounded the quotient up

Why? Sample answer: Mariana did not have enough money left over to buy another item for $17.

4 Jeremy wants to read 100 more books by the end of the school year. There are 36 weeks of school. How many books does Jeremy need to read each week?

SRB
44, 109, 113-114

Sample answer:

Number model: $\underline{100 \div 36 = b}$

Sample work:

$$36\overline{)100}$$
$$\underline{-\ 72} \quad 2 \qquad 28 \div 36 = \frac{28}{36}$$
$$28 \quad 2$$

Quotient: ___2___ Remainder: ___28___

Answer: Jeremy needs to read

_____ books each week. $2\frac{28}{36}$

Circle what you did with the remainder.

Ignored it

(Reported it as a fraction)

Rounded the quotient up

Why? Sample answer: It makes sense that Jeremy could read part of a book.

Math Boxes

Math Boxes

---

1 Insert grouping symbols to make true number sentences.

a. $4 * (8 - 5) = 12$

b. $2 + (7 * 7) = 51$

c. $91 / (4 - 3) = 91$

d. $[20 * (2 + 1) + 3] / 9 = 7$

e. $(60 + 12) / (30 + 6) = 2$

SRB
42-43

---

2 Complete the following equivalents.

a. 1 cup = __8__ ounces

b. 1 pint = __16__ ounces

c. 1 quart = __32__ ounces

d. 4 quarts = __128__ ounces

e. 1 gallon = __128__ ounces

SRB
215-217, 328

---

3 Which shows 672,891 in expanded form?

Fill in the circle next to the best answer.

(A) $6 \times 10,000 + 7 \times 1,000 + 2 \times 100 + 8 \times 10 + 9 \times 1 + 1 \times 0$

(B) 6 [100,000s] + 7 [1,000s] + 2 [100s] + 8 [10s] + 9 [1s] + 1

(C) $6 \times 10^5 + 7 \times 10^4 + 2 \times 10^3 + 8 \times 10^2 + 9 \times 10^1 + 1 \times 10^0$

(D) 670,000 + 2,800 + 90 + 1

SRB
70

---

4 What is the volume of the prism?

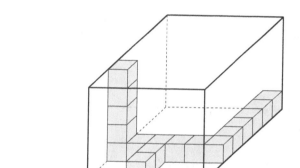

Volume = __240__ units^3

SRB
231-233

---

5 Solve mentally by breaking the dividend into smaller parts that are easier to divide. Write the equivalent name you used.

$840 \div 20 \rightarrow$ __42__

Equivalent name for 840:

Sample answer:

800 + 40

SRB
107

---

6 Solve. Use an estimate to check your answer.

$$\begin{array}{r} 1,\ 1\ 1\ 3 \\ \times \qquad 3\ 7 \\ \hline 4\ 1,\ 1\ 8\ 1 \end{array}$$

Answers vary.

(estimate)

SRB
83, 100-104

---

① 5.OA.1 ② 5.MD.1 ③ 5.NBT.1 ④ 5.MD.3, 5.MD.3b, 5.MD.4
⑤ 5.NBT.6 ⑥ 5.NBT.5

Fractions on a Number Line

1 Use the number line below for the Math Message. Sample answer:

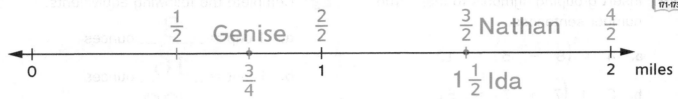

$\frac{1}{2}$ Genise $\frac{2}{2}$ $\frac{3}{2}$ Nathan $\frac{4}{2}$

0                    $\frac{3}{4}$ 1 $1\frac{1}{2}$ Ida   2   miles

2 Gary ran $1\frac{2}{3}$ miles and Lena ran $\frac{7}{6}$ mile. Who ran farther? Use the number lines below to help you answer the question.

  a. Divide this number line to show thirds. Label each tick mark. Then place a dot at $1\frac{2}{3}$.

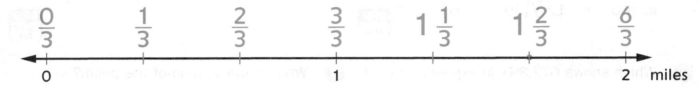

$\frac{0}{3}$ $\frac{1}{3}$ $\frac{2}{3}$ $\frac{3}{3}$ $1\frac{1}{3}$ $1\frac{2}{3}$ $\frac{6}{3}$

0   1   2   miles

  b. Divide this number line to show sixths. Label each tick mark. Place a dot at $\frac{7}{6}$.

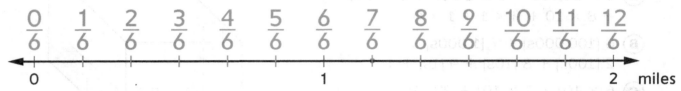

$\frac{0}{6}$ $\frac{1}{6}$ $\frac{2}{6}$ $\frac{3}{6}$ $\frac{4}{6}$ $\frac{5}{6}$ $\frac{6}{6}$ $\frac{7}{6}$ $\frac{8}{6}$ $\frac{9}{6}$ $\frac{10}{6}$ $\frac{11}{6}$ $\frac{12}{6}$

0   1   2   miles

  c. Who ran farther? _____Gary_____

3 Which number is greater? Circle the greater number in each pair. Use the Fraction Number Lines Poster or fraction circle pieces to help you.

  a. $\frac{5}{8}$ or $\left(\frac{9}{10}\right)$ **b.** $\frac{5}{3}$ or $\left(1\frac{5}{6}\right)$ **c.** $\left(2\frac{1}{4}\right)$ or $\frac{20}{12}$ **d.** $\left(\frac{9}{6}\right)$ or $\frac{13}{12}$

5.NF.2, 5.NF.3, SMP2, SMP5

4 Rachel and Dan are growing plants in science class. Rachel reports that her plant is $1\frac{1}{4}$ inches tall. Dan says his plant is $\frac{5}{2}$ inches tall.

a. Whose plant is taller? __Dan's plant__

b. How do you know? __Sample answer: $\frac{5}{2}$ is the same as__ $2\frac{1}{2}$ and $2\frac{1}{2}$ is greater than $1\frac{1}{4}$.

For Problems 5–10, rename each fraction as a mixed number or each mixed number as a fraction greater than 1. You may use the Fraction Number Lines Poster, fraction circle pieces, or division.

5 $\frac{5}{3} =$ __$1\frac{2}{3}$__

6 $1\frac{7}{9} =$ __$\frac{16}{9}$__

7 $\frac{11}{8} =$ __$1\frac{3}{8}$__

8 $1\frac{5}{6} =$ __$\frac{11}{6}$__

9 $\frac{16}{5} =$ __$3\frac{1}{5}$__

10 $2\frac{1}{3} =$ __$\frac{7}{3}$__

Try This

11 a. Rename $\frac{34}{8}$ as a mixed number. __$4\frac{2}{8}$__

b. Explain your reasoning.
__Sample answer: $\frac{34}{8} = 34 \div 8$, which is 4 R2, or $4\frac{2}{8}$.__

Math Boxes

1 Solve.

a. $3 * 10^1 =$ ___30___

b. $3 * 10^2 =$ ___300___

c. $3 * 10^3 =$ ___3,000___

d. $3 * 10^4 =$ ___30,000___

Write a number sentence that follows the pattern above.

___3___ * ___10^5___ = ___300,000___

SRB
95-96

2 Solve.

$32\overline{)6,572}$

$6,572 \div 32 \rightarrow$ ___205 R12___

SRB
108-110

3 What is the area of a garden plot that measures $6\frac{1}{2}$ feet by 4 feet?

4 ft

$6\frac{1}{2}$ ft

Area = ___26___ square feet

Sample answer:

(number model)

$6\frac{1}{2} * 4 = 26$

SRB
224-225

4 Jamar bought eight 6-packs of juice boxes for his family. His grandmother bought 3 more 6-packs. Write an expression that models how many juice boxes they bought.

Sample answer:

$6 * (8 + 3)$

SRB
44

5 **Writing and Reasoning** Ari made a sketch to solve Problem 2. Use Ari's sketch to explain how you think he solved the problem.

Area: 6,572

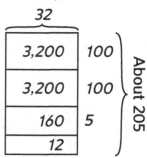

Sample answer: Ari used an area model. He used the partial quotients 100, 100, and 5. There was a remainder of 12. This gave him an answer of 205 R12.

SRB
8-9,
111-112

① 5.NBT.2 ② 5.NBT.6 ③ 5.NF.4, 5.NF.4b ④ 5.OA.1, 5.OA.2
⑤ 5.NBT.6, SMP2, SMP3

Harjit is playing a version of *Division Top-It* with a friend. In this version each player turns over 3 number cards and places them as the digits in the division problem below. The player with the larger quotient wins the round.

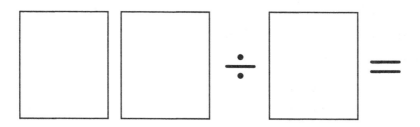

Harjit turns over a 6, a 3, and a 9. How do you think Harjit should place her cards to get the largest possible quotient? Explain your thinking.

Sample answer: Harjit should use her cards to make the problem 96 ÷ 3 = 32. 96 is the largest possible 2-digit number and 3 is the smallest possible divisor. I know that the larger the dividend, the more there is to divide up, so the larger the quotient. I also know that the smaller the divisor, the larger the quotient. So in order to get the largest quotient, Harjit should make the largest dividend and the smallest divisor.

Math Boxes

1 Which number stories have the answer $\frac{3}{4}$? Circle ALL that apply.

A. Four dogs shared 3 dog treats. How many treats did each dog get?

B. Three friends shared 4 oranges. How much orange did each friend eat?

C. Four dogs each ate 3 dog treats. How many treats did the dogs eat?

D. Four friends shared 3 oranges. How much orange did each friend eat?

SRB
163-164

2 Estimate and fill in the missing digits.

Answers vary.
_____
(estimate)

```
              9   7
    ×         7   6
          5  [8]  2
    + [6]  7 [9]  0
    [7],[3]  7 [2]
```

SRB
83, 103

3 Complete the table.

Exponential Notation	Standard Notation
10^0	1
10^1	10
10^2	100
10^5	100,000
10^7	10,000,000

SRB
68

4 Solve.

a. $63 / 9 =$ **7**

b. $630 / 9 =$ **70**

c. $630 / 90 =$ **7**

d. $6,300 / 900 =$ **7**

e. $63,000 / 9 =$ **7,000**

SRB
106

5 Find the volume of a box with a base area of 24 in.² and a height of 12 in. Use the formula $V = B \times h$.

Volume = **24** × **12**

Volume = **288** in.³

SRB
233

6 Place the fractions on the number line.

$\frac{1}{2}$ $\frac{3}{4}$ $\frac{1}{4}$

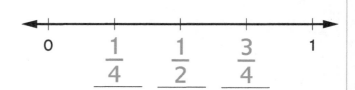

SRB
158, 161

① 5.NF.3 ② 5.NBT.5 ③ 5.NBT.2 ④ 5.NBT.6
84 ⑤ 5.MD.5, 5.MD.5b ⑥ 5.NF.3

Christopher solved some fraction problems. Do Christopher's answers make sense? Circle Yes or No. Then write an argument to show how you know.

SRB
10-11,
181-185

Name __Christopher__

Sample arguments given.

Solve.

① $\frac{2}{7} + \frac{1}{2} = \frac{3}{9}$

Conjecture: Does answer 1 make sense? **Yes** (**No**)

Argument: If you add $\frac{2}{7}$ to $\frac{1}{2}$, the answer has to be more than $\frac{1}{2}$, not less. The answer $\frac{3}{9}$ is less than $\frac{1}{2}$ so it can't be correct.

---

② Write >, <, or =.

$\frac{9}{10} \ \underline{\ >\ } \ \frac{7}{8}$

Conjecture: Does answer 2 make sense? (**Yes**) **No**

Argument: I compared each fraction to 1. $\frac{9}{10}$ is $\frac{1}{10}$ away from 1. $\frac{7}{8}$ is $\frac{1}{8}$ away from 1. $\frac{1}{10}$ is less than $\frac{1}{8}$, so I know that $\frac{9}{10}$ is closer to 1 whole. $\frac{9}{10}$ is greater than $\frac{7}{8}$.

---

③ $\frac{7}{12} + \frac{1}{4} = \frac{8}{12}$

Conjecture: Does answer 3 make sense? **Yes** (**No**)

Argument: $\frac{7}{12}$ and $\frac{8}{12}$ have the same denominator so they're easy to compare. I know that I would have to add $\frac{1}{12}$ to $\frac{7}{12}$ to get $\frac{8}{12}$. $\frac{1}{4}$ does not equal $\frac{1}{12}$, so the answer doesn't make sense.

---

④ $\frac{8}{9} + \frac{1}{3} = \frac{8}{12}$

Conjecture: Does answer 4 make sense? **Yes** (**No**)

Argument: I compared the denominators because $\frac{8}{12}$ and $\frac{8}{9}$ have the same numerator. 12ths are smaller than 9ths, so $\frac{8}{12} < \frac{8}{9}$. It doesn't make sense for the sum to be smaller than the addend.

Interpreting Remainders in Number Stories

For each number story, write a number model with a letter for the unknown. Then solve. Show your work in the space provided. You can draw a picture to help. Decide what to do with the remainder and explain what you did. Sample number models, explanations, and work given.

SRB 44, 108, 113-114

1 A cook has 250 ounces of cheese for 80 individual pizzas. Each pizza gets the same amount of cheese. How much cheese should the cook put on each pizza?

Number model: __250 ÷ 80 = c__

Quotient: __3__   Remainder: __10__

Answer: The cook should put _____ ounces of cheese on each pizza. $3\frac{10}{80}$, or $3\frac{1}{8}$

Circle what you did with the remainder.

$8 * 3 = 24$

$80 * 3 = 240$

So $250 ÷ 80 \rightarrow$

3 R10

$10 ÷ 80 = \frac{10}{80}$, or $\frac{1}{8}$

Ignored it   ⟨Reported it as a fraction⟩   Rounded the quotient up

Why? __It doesn't make sense to waste the extra cheese,__ __so I split up the 10 ounces left over into fractions of an__ __ounce.__

2 35 people are attending a game night. Each table seats 4 people. How many tables are needed?

Number model: __35 ÷ 4 = t__

Quotient: __8__   Remainder: __3__

Answer: __9__ tables are needed.

Circle what you did with the remainder.

How many tables?
4 people per table

Ignored it   Reported it as a fraction   ⟨Rounded the quotient up⟩

Why? __Another table is needed to seat the 3 people left over.__

Math Boxes

1 Liz bought four 2-quart packages of strawberries. How many gallons of strawberries did she buy?

$(4 × 2) / 4 = g$

(number model)

Answer: __2__ gallons

SRB
44, 214-
217, 328

2 Liliana is packing 135 winter hats into donation boxes. 32 hats will fit in a box. How many boxes does she need?

$135 ÷ 32 = b$

(number model)

Quotient: __4__ Remainder: __7__

Answer: She needs __5__ boxes.

SRB
44, 109,
113

3 Four friends equally shared 6 cups of soup. How much soup did each friend get? Circle ALL that apply.

(A.) $\frac{6}{4}$ cups

(B.) $1\frac{2}{4}$ cups

(C.) $1\frac{1}{2}$ cups

D. $\frac{4}{6}$ cup

SRB
163-164,
170-171

4 Complete. Sample answers:

a. Write a number in which a 5 is worth 500. __8,521__

Write a number in which a 5 is worth 10 times as much. __85,328__

b. Write a number in which a 7 is worth 70,000. __72,987__

Write a number in which a 7 is worth $\frac{1}{10}$ as much. __7,999__

SRB
66-67

5 **Writing/Reasoning** What did you decide to do with the remainder in Problem 2? Why? Sample answer: I decided to round the quotient up because there were 7 hats that wouldn't fit in the first 4 boxes. Liliana would need one more box to carry the 7 extra hats, so the answer is 5 boxes.

SRB
113

① 5.MD.1 ② 5.NBT.6 ③ 5.NF.3 ④ 5.NBT.1
⑤ 5.NBT.6, SMP1, SMP6   87

Using Benchmarks to Make Estimates

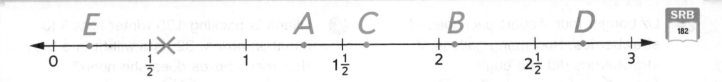

SRB
182

Estimate the sum or difference for each fraction number story. Place an X on the number line to represent your estimate. Then circle the best answer.

1. Micah bought $1\frac{1}{3}$ pounds of grapes and $1\frac{1}{2}$ pounds of bananas. About how many pounds of fruit did Micah buy?

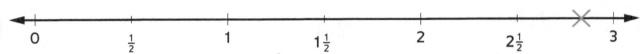

How much fruit?   about 2 pounds   about $2\frac{1}{2}$ pounds   (about 3 pounds)

Explain your thinking.

Sample answer: $1\frac{1}{3}$ is close to $1\frac{1}{2}$, so I added $1\frac{1}{2} + 1\frac{1}{2}$ in my head. $1 + 1 = 2$, and $\frac{1}{2} + \frac{1}{2}$ is 1, so that's 3 all together.

2. Chloe has $2\frac{1}{2}$ yards of fabric. She will use about $\frac{3}{8}$ yard to make a scarf. How many yards of fabric will she have left?

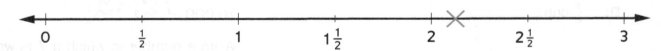

How much fabric is left?   about $1\frac{1}{2}$ yards   (about 2 yards)   about 3 yards

Explain your thinking.

Sample answer: I used a benchmark of $\frac{1}{2}$ for $\frac{3}{8}$. Then I thought about $2\frac{1}{2} - \frac{1}{2}$ and knew that would be about 2.

Try This

3. The perimeter of a triangle is 10 inches. One side is $3\frac{9}{16}$ inches long. Another side is $4\frac{5}{8}$ inches long. About how many inches long is the third side? Explain how you estimated.

Sample answer: $3\frac{9}{16}$ is close to $3\frac{1}{2}$, and $4\frac{5}{8}$ is close to $4\frac{1}{2}$.

The length of those two sides is about 8 inches.

So the third side is about $10 - 8$, or about 2 inches.

Math Boxes

1 Write a division number story with an answer of $\frac{3}{5}$.   Sample answer:

5 people shared 3 apples

equally. How much apple

did each person get?

SRB
163-164

2 Estimate. Fill in the missing digits.

Estimate: ___Answers vary.___

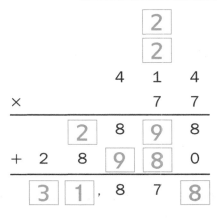

```
            2
            2
      4   1   4
  ×         7   7
      2   8   9   8
  +  2   8   9   8   0
      3   1 ,  8   7   8
```

SRB
83, 103

3 Rewrite each number in standard or exponential notation.

a. 10^3 = ___1,000___

b. 10,000 = ___10^4___

c. 10^5 = ___100,000___

d. 1,000,000 = ___10^6___

e. 10^8 = ___100,000,000___

SRB
68

4 Solve.

How many 8s in 72? ___9___

How many 800s in 72,000? ___90___

How many 5s in 450,000? ___90,000___

How many 3,000s in 270,000? ___90___

How many 90s in 63,000? ___700___

SRB
106

5 Find the volume of a shipping container that is 20 feet long, 8 feet wide, and 8 feet tall. Use the formula $V = l \times w \times h$.

V = ___20___ × ___8___ × ___8___

Volume = ___1,280___ ft³

SRB
233

6 Place the fractions on the number line.

$\frac{1}{3}$ $\frac{2}{3}$ $\frac{4}{3}$ $\frac{5}{3}$

0    $\frac{1}{3}$ $\frac{2}{3}$ 1 $\frac{4}{3}$ $\frac{5}{3}$ 2

SRB
158-161

① 5.NF.3 ② 5.NBT.5 ③ 5.NBT.2 ④ 5.NBT.6
⑤ 5.MD.5, 5.MD.5b ⑥ 5.NF.3

89

Math Boxes

Renaming Fractions and Mixed Numbers

Solve by following the steps. You can use fraction circles, number lines, or drawings to help.

SRB
171-173

1 Find another name for $2\frac{3}{4}$.

- Show $2\frac{3}{4}$.

- Break apart 1 whole into $\frac{4}{4}$.

Name: $1\frac{7}{4}$

2 Find another name for $\frac{14}{3}$.

- Show $\frac{14}{3}$.

- Make as many groups of 3 thirds as you can.

- Trade each $\frac{3}{3}$ for 1 whole.

Name: $4\frac{2}{3}$

Write another name for each mixed number that has the same denominator. Check that your trades are fair and record them. Sample answers given for Problems 3, 4, and 8.

Example: $3\frac{8}{6}$

Name: $2\frac{14}{6}$

Trade: *1 whole for $\frac{6}{6}$*

3 $2\frac{4}{5}$

Name: $1\frac{9}{5}$

Trade: 1 whole for $\frac{5}{5}$

4 $1\frac{12}{10}$

Name: $2\frac{2}{10}$

Trade: $\frac{10}{10}$ for 1 whole

Fill in the missing whole number or missing numerator.

5 $1\frac{4}{3} = 2\frac{\boxed{1}}{3}$

6 $\boxed{3}\frac{7}{5} = 4\frac{2}{5}$

7 $2\frac{9}{2} = 4\frac{\boxed{5}}{2}$

8 **a.** Mojo the monkey has 2 whole bananas and 5 half-bananas. Write a mixed number to show how many bananas Mojo has. $2\frac{5}{2}$ bananas

b. Manny the monkey has 4 whole bananas and 1 half-banana. Do Mojo and Manny have the same amount of banana? Explain how you know. Yes. If Mojo traded 4 of his half-bananas he would have 4 whole bananas and 1 half-banana. That's the same as Manny.

c. Marcus the monkey has the same amount of banana as Mojo. He only has half-bananas. How many half-bananas does he have? Explain your answer. He has 9 half-bananas, because $\frac{9}{2}$ is another name for $2\frac{5}{2}$.

5.NF.3, SMP3, SMP5

Connecting Fractions and Division

Write a division expression to model each story, then solve. You can use fraction circles or draw pictures to help.

1 Olivia is running a 3-mile relay race with 3 friends. If the 4 of them each run the same distance, how many miles will each person run?

Division number model: _____ $3 \div 4$ _____

Fractional answer: _____ $\frac{3}{4}$ mile _____

2 Chris has 3 pints of blueberries for fruit salad. If he splits the blueberries equally into 8 serving bowls, how many pints of blueberries will be in each bowl?

Division number model: _____ $3 \div 8$ _____

Fractional answer: _____ $\frac{3}{8}$ pint _____

3 Four students shared 9 packages of pencils equally. How many packages of pencils did each student get?

Division number model: _____ $9 \div 4$ _____

Fractional answer: _____ $\frac{9}{4}$, or $2\frac{1}{4}$, packages _____

Use your answers to Problems 1–3 to answer the questions below.

4 Compare the numbers in each division number model with your fractional answer. What do you notice?

Sample answer: The dividend is the same as the answer's numerator. The divisor is the same as the answer's denominator.

5 Write a rule for finding the fractional answer to a division problem by using the dividend and divisor.

Sample answer: Make the dividend the numerator and the divisor the denominator.

6 Write and solve your own number story using your rule.

Sample answer: Six people split 5 apples equally. How much apple did each person get? Each person got $\frac{5}{6}$ apple.

5.OA.2, 5.NF.3, SMP7, SMP8

Math Boxes

1 Noelle bought two $\frac{1}{2}$-gallon cartons of lemonade for her lemonade stand. How many ounces of lemonade did she buy? Sample answer:

$$\left(\frac{1}{2} + \frac{1}{2}\right) * 128 = o$$

(number model)

Answer: _____128_____ ounces

SRB
214-217,
328

2 The student council has $289 to spend on decorations for the fall festival. If decorations for each table cost $13, how many tables can they decorate?

$$289 \div 13 = b$$

(number model)

Quotient: _____22_____ Remainder: _____3_____

Answer: They can decorate _____22_____ tables.

SRB
44, 109,
113

3 The fifth-grade teachers ordered 3 boxes of clay. They wanted to share it equally among their four classes. How much clay does each class get?

Division number model:

$$3 \div 4 = c$$

Answer: _____$\frac{3}{4}$_____ box of clay

SRB
163-164

4 Move the digits in 625,134 to create a new number.

Move the 2 so it is worth $\frac{1}{10}$ as much.

Move the 3 so it is worth 10 times as much.

Move the 5 so it is worth 50,000.

Move the 4 so its value changes to $4 \times 100,000$.

Move the 1 and the 6 so that the sum of their values is 16.

Write the new number:

_____452,316_____

SRB
66-67

5 **Writing/Reasoning** Draw a picture to show how you solved Problem 3.

Sample answer:

1	2
3	4

1	2
3	4

1	2
3	4

SRB
12-14,
163-164

① 5.MD.1 ② 5.NBT.6 ③ 5.NF.3 ④ 5.NBT.1
⑤ 5.NF.3, SMP4

92

Addition and Subtraction Number Stories

SRB
178-185

For each story: Sample number models, estimates, and work are given.
- Write a number model with a letter for the unknown.
- Make an estimate.
- Solve. You can use fraction circle pieces, a drawing, or a number line to help.
- Use your estimate to check whether your answer makes sense.

1 Andrea had $1\frac{1}{5}$ liters of water. She drank $\frac{3}{5}$ liter. How much did she have left?

Number model: $1\frac{1}{5} - \frac{3}{5} = w$

Estimate: Less than 1 liter

Answer: $\frac{3}{5}$ liter

2 A table is $2\frac{8}{12}$ feet tall and a lamp on it is $1\frac{5}{12}$ feet tall. What is their total height?

Number model: $2\frac{8}{12} + 1\frac{5}{12} = h$

Estimate: Between 4 and $4\frac{1}{2}$ feet

Answer: $4\frac{1}{12}$ feet

3 A chef had $2\frac{5}{8}$ pitas. She used $1\frac{7}{8}$ pitas. How many pitas does she have left?

Number model: $2\frac{5}{8} - 1\frac{7}{8} = p$

Estimate: About $\frac{5}{8}$ pita

See sample answer for Problem 5 for sample strategy.

Answer: $\frac{6}{8}$ pita

4 Niko rode a bike $2\frac{3}{10}$ miles. Then he rode another $2\frac{8}{10}$ miles. How far did he ride?

Number model: $2\frac{3}{10} + 2\frac{8}{10}$

Estimate: Between 5 and $5\frac{1}{2}$ miles

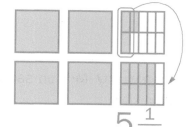

Answer: $5\frac{1}{10}$ miles

5 Explain how you solved Problem 3. Sample answer: I showed $2\frac{5}{8}$ with fraction circle pieces. Then I traded one whole for $\frac{8}{8}$ because I didn't have enough eighths to take away. So then I had $1\frac{13}{8}$. I took away $1\frac{7}{8}$ and I had $\frac{6}{8}$ left.

5.NF.2, SMP1, SMP4, SMP5

93

Math Boxes

1 Allison was making pancakes. She needed $\frac{1}{4}$ cup of vegetable oil and $\frac{3}{4}$ cup of milk. How many cups of liquid did she need?

$$\frac{1}{4} + \frac{3}{4} = c$$
_____
(number model)

Answer: $\frac{4}{4}$, or 1 _____ cup(s) of liquid

SRB
178-180,
186

2 Solve.

a. $\begin{array}{r} 2\ 2\ 5 \\ \times \quad\quad 2 \\ \hline 4\ 5\ 0 \end{array}$ b. $\begin{array}{r} 2\ 2\ 5 \\ \times \quad\quad 4 \\ \hline 9\ 0\ 0 \end{array}$

SRB
100,
102, 104

3 Solve.

a. $(4 * 12) + 8 =$ ___56___

b. ___1___ $= (32 / 16) / 2$

c. $(65 + 83) / (3 - 1) =$ ___74___

d. ___7___ $= 3 + \{32 / (16 / 2)\}$

SRB
42-43

4 Write fractions that make the number sentences true. Sample answers:

a. $\dfrac{\frac{1}{2}}{} + \dfrac{\frac{1}{6}}{} < 1$

b. $\dfrac{\frac{5}{6}}{} - \dfrac{\frac{1}{4}}{} < 1$

c. $\dfrac{\frac{2}{3}}{} + \dfrac{\frac{3}{4}}{} > 2$

d. $\dfrac{\frac{1}{2}}{} + \dfrac{\frac{1}{4}}{} < 1\frac{1}{2}$

SRB
181-184

5 **Writing/Reasoning** Morton says he can compare the products in Problem 2 without multiplying. Explain how.

Sample answer: Morton knows the product of 225 × 4 is going to be 2 times as large as the product of 225 × 2 because 4 is 2 times as large as 2.
_____

_____

SRB
46

① 5.NF.2 ② 5.NBT.5 ③ 5.OA.1 ④ 5.NF.2
94   ⑤ 5.OA.2, SMP6, SMP7

Adding Fractions with Circle Pieces

Make an estimate. Then use your fraction circle pieces to find the sum. Use the red circle as the whole. Remember to think about using same-size pieces.

SRB 155, 189

Write a number sentence to show how you used equivalent fractions to find the sum.

Example: $\frac{1}{2} + \frac{1}{8} = ?$

Estimate: _Between $\frac{1}{2}$ and 1_

Show $\frac{1}{2} + \frac{1}{8}$ with fraction pieces.

Cover the $\frac{1}{2}$ piece with four $\frac{1}{8}$ pieces to show that $\frac{1}{2}$ is the same as $\frac{4}{8}$.

Sum: $\frac{5}{8}$

Number sentence: $\frac{4}{8} + \frac{1}{8} = \frac{5}{8}$

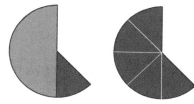

Sample estimates are given.

1 $\frac{2}{3} + \frac{1}{6} = ?$

Estimate: A little less than 1

Sum: $\frac{5}{6}$

Number sentence: $\frac{4}{6} + \frac{1}{6} = \frac{5}{6}$

2 $\frac{2}{5} + \frac{3}{10} = ?$

Estimate: Between $\frac{1}{2}$ and 1

Sum: $\frac{7}{10}$

Number sentence: $\frac{4}{10} + \frac{3}{10} = \frac{7}{10}$

3 $\frac{1}{3} + \frac{1}{12} = ?$

Estimate: About $\frac{1}{2}$

Sum: $\frac{5}{12}$

Number sentence: $\frac{4}{12} + \frac{1}{12} = \frac{5}{12}$

4 $\frac{2}{6} + \frac{1}{4} = ?$

Estimate: A little more than $\frac{1}{2}$

Sum: $\frac{7}{12}$

Number sentence: $\frac{4}{12} + \frac{3}{12} = \frac{7}{12}$

5 $\frac{2}{3} + \frac{1}{4} = ?$

Estimate: Close to 1

Sum: $\frac{11}{12}$

Number sentence: $\frac{8}{12} + \frac{3}{12} = \frac{11}{12}$

6 $\frac{1}{2} + \frac{1}{5} = ?$

Estimate: Between $\frac{1}{2}$ and 1

Sum: $\frac{7}{10}$

Number sentence: $\frac{5}{10} + \frac{2}{10} = \frac{7}{10}$

Explaining Place-Value Patterns

1 Solve.

   a. 45 * 10 = _____ 450 _____

   b. 45 * 10 * 10 = _____ 4,500 _____

   c. 45 * 10 * 10 * 10 = _____ 45,000 _____

   d. 45 * 10 * 10 * 10 * 10 = _____ 450,000 _____

2 Look at your answers to Problem 1.

   a. What pattern do you notice in the number of zeros?

   Sample answer: Every time I multiply by one more 10,
there is one more zero in the answer.

   b. What pattern do you notice in the value of the products?

   Sample answer: Every time I multiply by one more 10
the product is 10 times as large.

   c. Do you think the patterns will hold true no matter how many 10s are in the problem?
Use what you know about place value to explain your answer.

   Sample answer: The pattern will keep going. Every time I
multiply by 10, I write one more zero. That means the
product will be 10 times as much because all the digits
move one place-value position to the left.

5.NBT.1, 5.NBT.2, SMP7, SMP8

SRB
68-69.
95-96

3 Solve.

a. $328 * 10^2 = $ _____32,800_____

b. $328 * 10^5 = $ _____32,800,000_____

c. $328 * 10^7 = $ _____3,280,000,000_____

d. $328 * 10^4 = $ _____3,280,000_____

e. $328 * 10^3 = $ _____328,000_____

f. $328 * 10^1 = $ _____3,280_____

4 Look at your answers to Problem 3.

a. What pattern do you notice in the number of zeros?

Sample answer: The number of zeros at the end of each answer is the same as the exponent.

b. Use the pattern to help you write a rule for how to multiply a whole number by a power of 10.

Sample answer: To multiply a whole number by a power of 10, first write the whole number. Then write the number of zeros shown by the exponent.

c. Use what you know about place value to explain why your rule will always work.

Sample answer: The exponent shows how many times you are multiplying by 10. Every time you multiply by 10, you just write one more zero because multiplying by 10 moves all the digits one place to the left.

Math Boxes: Preview for Unit 4

Math Boxes

---

① Write each number in standard notation.

a. $(3 \times 1{,}000{,}000) + (4 \times 100{,}000) +$
$(2 \times 10{,}000) + (1 \times 1{,}000) +$
$(9 \times 100) + (8 \times 10) + (7 \times 1) =$
3,421,987

b. ___82,456___ $= 8\ [10{,}000s] +$
$2\ [1{,}000s] + 4\ [100s] + 5\ [10s] +$
$6\ [1s]$

c. $100{,}000 + 20{,}000 + 8{,}000 + 300 +$
$20 + 8 =$ __128,328__

SRB 70

---

② **a.** Round 42 to the nearest ten.
40

b. Round 382 to the nearest hundred.
400

c. Round 8,461 to the nearest thousand.
8,000

d. Round 4.2 to the nearest whole.
4

SRB 79-82, 126-127

---

③ Write the money amounts in dollars-and-cents notation.

one dollar and ten cents
$1.10

three dollars and fifty-two cents
$3.52

Circle the amount that is greater.

(**$5.75**) $5.57

SRB 121-123   SRB 275

---

④ Place the numbers 3, 6, 9, 2, and 4 on the number line.

---

⑤ Write each fraction as a decimal.

a. $\frac{4}{10}$ ___0.4___

b. $\frac{8}{10}$ ___0.8___

c. $\frac{52}{100}$ ___0.52___

d. $\frac{40}{100}$ ___0.40___

SRB 116

---

⑥ What is the value of the bold digit?

a. $5.\mathbf{4}3$ ___40 cents___

b. $\mathbf{6}.27$ ___6 dollars___

c. $\mathbf{8}2.76$ ___80 dollars___

d. $9.0\mathbf{2}$ ___2 cents___

SRB 118-119

---

① 5.NBT.1, 5.NBT.3, 5.NBT.3a ② 5.NBT.4 ③ 5.NBT.3, 5.NBT.3a, 5.NBT.3b ④ 5.G.2 ⑤ 5.NBT.3, 5.NBT.3a

⑥ 5.NBT.3, 5.NBT.3a, 5.NBT.3b

Renaming Fractions and Mixed Numbers

Fractions greater than 1 can be expressed as mixed numbers, such as $2\frac{1}{3}$ and $1\frac{4}{3}$, and as fractions with a numerator larger than the denominator, such as $\frac{7}{3}$. You know several ways to rename fractions as mixed numbers and mixed numbers as fractions.

SRB
171-173

Use fraction circle pieces: Show the original number. Make fair trades between wholes and same-size pieces to rename.

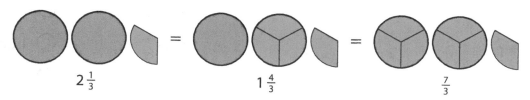

$2\frac{1}{3}$ $1\frac{4}{3}$ $\frac{7}{3}$

Use number lines: Use fraction names for whole numbers and count up. The number line on the left shows that $2\frac{1}{3}$ is $\frac{4}{3}$ past 1, so $2\frac{1}{3} = 1\frac{4}{3}$. The number line on the right shows that $2 = \frac{6}{3}$, and $2\frac{1}{3}$ and $\frac{7}{3}$ are both $\frac{1}{3}$ past 2, so $2\frac{1}{3} = \frac{7}{3}$.

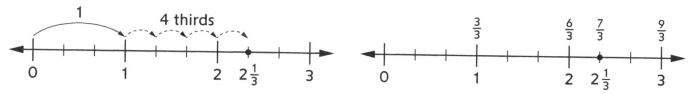

Think about making or breaking wholes:

• To rename $2\frac{1}{3}$ as a fraction, think: *How many thirds are in $2\frac{1}{3}$?* There are 2 wholes. I can break each whole into 3 thirds. Two groups of 3 thirds is the same as 2 * 3 thirds = 6 thirds, or $\frac{6}{3}$. Add one more third to get $\frac{7}{3}$.

• To rename $\frac{7}{3}$ as a mixed number, think: *How many groups of 3 thirds are in 7? What's left over?* There are 2 groups of 3 thirds in 7, with 1 third left over. $\frac{7}{3} = 2\frac{1}{3}$

For Problems 1–3, rename each fraction as a mixed number. Make as many wholes as you can.

① $\frac{9}{4} =$ ___$2\frac{1}{4}$___

② $\frac{12}{5} =$ ___$2\frac{2}{5}$___

③ $\frac{15}{8} =$ ___$1\frac{7}{8}$___

For Problems 4–6, write at least two other names with the same denominator for each mixed number.

④ $3\frac{4}{5} =$ ___$2\frac{9}{5}, 1\frac{14}{5}, \frac{19}{5}$___

⑤ $2\frac{1}{6} =$ ___$1\frac{7}{6}, \frac{13}{6}$___

⑥ $4\frac{1}{2} =$ ___$3\frac{3}{2}, 2\frac{5}{2}, 1\frac{7}{2}, \frac{9}{2}$___

For Problems 7–9, fill in the missing whole number or missing numerator.

⑦ $4\frac{2}{5} = \frac{22}{5}$

⑧ $2\frac{2}{8} = \frac{18}{8}$

⑨ $2\frac{5}{3} = 3\frac{2}{3}$

5.NF.3, SMP2, SMP5

99

Math Boxes

1 Estimate and solve.

<u> Answers vary. </u>

(estimate)

$32\overline{)728}$

$728 \div 32 \rightarrow$ __22 R24__

SRB 84, 109-110

2 Use estimation strategies to determine whether the number sentences are true or false.

a. $\frac{3}{4} + \frac{1}{2} < 1$ ○ True   ● False

b. $1 - \frac{3}{4} > \frac{1}{2}$ ○ True   ● False

c. $\frac{2}{3} + \frac{1}{8} > \frac{1}{2}$ ● True   ○ False

d. $\frac{3}{2} + \frac{2}{7} > 1\frac{1}{2}$ ● True   ○ False

SRB 181-182

3 Solve. Use fraction circle pieces to help you.

a. $\frac{1}{2} + \frac{1}{4} =$ __$\frac{3}{4}$__

b. $\frac{1}{2} + \frac{2}{6} =$ __$\frac{5}{6}$__

c. $\frac{4}{8} + \frac{1}{2} =$ __$\frac{8}{8}$, or 1__

d. $\frac{2}{3} + \frac{1}{6} =$ __$\frac{5}{6}$__

SRB 166, 189

4 Sophia bought 3 flashlights. Each flashlight cost 5 dollars. The batteries for each flashlight cost 2 dollars. Which expression models this situation?

Fill in the circle next to the best answer.

Ⓐ $(3 * 5) + 2$ Ⓑ $(3 + 2) * 5$

Ⓒ $(5 + 2) / 3$ **Ⓓ** $(5 + 2) * 3$

SRB 42, 44

5 A cook needs to know the volume of his cupboards. Use the model of the cupboards shown below to estimate the total volume of the cupboards.

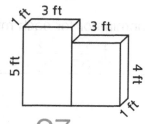

Volume = __27__ cubic feet

SRB 233-234

6 There are 24 classes at Lincoln School, each with 23 students.
Write an expression to model how many students go to the school. Then evaluate the expression to find the number of students at the school.

Expression: __24×23__

Answer: __552__ students

SRB 38, 44, 100-104

① 5.NBT.6 ② 5.NF.2 ③ 5.NF.1 ④ 5.OA.1, 5.OA.2
⑤ 5.MD.5, 5.MD.5b, 5.MD.5c ⑥ 5.OA.2, 5.NBT.5

Solving Fraction Number Stories

Solve each number story. You can use fraction circle pieces, number lines, pictures, and other tools or models to help you. Show your work.

SRB
30,
178-181

1 Four friends shared 5 sandwiches after their hike. If they each ate an equal share, how many sandwiches did each friend eat?

Sample work:

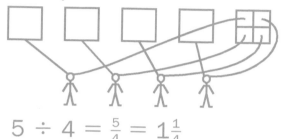

$5 \div 4 = \frac{5}{4} = 1\frac{1}{4}$

Answer: Each friend ate $\frac{5}{4}$, or $1\frac{1}{4}$, sandwiches.

2 Josh combined $\frac{1}{2}$ carton of eggs with $\frac{1}{3}$ carton of eggs. He said, "Now I have $\frac{2}{5}$ of a carton." Is Josh correct?

Answer: No.

How do you know? Sample answer: I know because $\frac{1}{2} + \frac{1}{3}$ should be more than $\frac{1}{2}$. When I used fraction circle pieces, I saw that 2 fifth pieces are less than 1 half piece.

3 Ryan lives $3\frac{1}{4}$ miles from school. Kayla lives $2\frac{3}{4}$ miles from school. How much farther from school does Ryan live than Kayla?

Sample work:

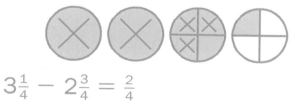

$3\frac{1}{4} - 2\frac{3}{4} = \frac{2}{4}$

Answer: Ryan lives $\frac{2}{4}$ mile farther from school than Kayla.

4 Delilah was playing *Fraction Capture*. She wrote her initials on a $\frac{1}{3}$ section and a $\frac{1}{6}$ section. What is the sum of the sections she initialed?

Sample work:

$\frac{1}{3} + \frac{1}{6} = ?$ $\frac{1}{3} = \frac{2}{6}$

$\frac{2}{6} + \frac{1}{6} = \frac{3}{6}$

Answer: The sum of Delilah's sections is $\frac{3}{6}$.

5.NF.1, 5.NF.2, 5.NF.3, SMP1, SMP4

SRB
30,
178-181

5 Lauren had $\frac{3}{4}$ gallon of paint. She poured in an additional $\frac{1}{8}$ gallon from another can. How much paint does Lauren have?

Sample work:

$\frac{3}{4} + \frac{1}{8} = ?$ $\frac{3}{4} = \frac{6}{8}$

$\frac{6}{8} + \frac{1}{8} = \frac{7}{8}$

Answer: She has $\frac{7}{8}$ gallon of paint.

6 Codyone is cutting fabric for a quilt. She has 4 feet of cloth to make 12 quilt pieces. If she uses all of the fabric, how long should each quilt piece be?

Sample work:

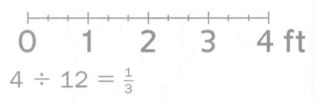

$4 \div 12 = \frac{1}{3}$

Answer: Each quilt piece should be $\frac{4}{12}$, or $\frac{1}{3}$, foot long.

7 Alyssa started with $\frac{7}{8}$ jar of jam. She used about $\frac{1}{16}$ of the jam to make a sandwich. Alyssa said, "I don't need to put jam on the grocery list yet. We still have about $\frac{3}{4}$ jar." Is Alyssa correct?

Answer: Yes.

How do you know?

Sample answer: I know that $\frac{6}{8}$ is the same as $\frac{3}{4}$. If Alyssa had $\frac{7}{8}$ jar and used $\frac{1}{16}$ of it, she still has more than $\frac{6}{8}$. So she still has plenty of jam.

8 Tyrell and his mom went grocery shopping. They bought $1\frac{1}{6}$ pounds of carrots and $2\frac{5}{6}$ pounds of potatoes. Tyrell carried the carrots and potatoes in one grocery bag. How heavy was the bag?

Sample work:

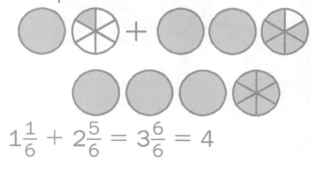

$1\frac{1}{6} + 2\frac{5}{6} = 3\frac{6}{6} = 4$

Answer: The bag weighed 4 pounds.

Practicing Division

For Problems 1 and 2, make an estimate.
Then divide using partial-quotients division. Report your remainder as a fraction.
Use your estimate to check that your answers make sense.

SRB
108-110,
113-114

① 5,926 / 48 = ? Sample answer:

Estimate: 6,000 / 50 = 120

② 9,031 / 71 = ? Sample answer:

Estimate: 9,000 / 90 = 100

Answer: $123\frac{22}{48}$

Answer: $127\frac{14}{71}$

For Problems 3 and 4, write a number model for the story using a letter for the unknown. Then solve the story. Remember to think about what you should do with the remainder.

③ A food pantry received a donation of 248 cans of chicken. They want to distribute them evenly to 9 different soup kitchens in the city. How many cans of chicken will each soup kitchen get?

Number model: 248 / 9 = c

④ Calvin is cutting a roll of paper into 4 pieces to make signs for the school carnival. The paper is 145 inches long. How long should he make each banner?

Number model: $145 \div 4 = b$

Answer: 27 cans

Answer: $36\frac{1}{4}$ inches

⑤ Explain how you found your answer to Problem 4. Sample answer:
I used partial-quotients division and found that 36 groups of 4 are in 145, with 1 left over. I divided the 1 into 4 equal parts since paper can be cut into fractions of an inch.

5.NBT.6, SMP1, SMP4

103

Math Boxes

1 Jackson had $\frac{5}{6}$ yard of ribbon. He used $\frac{2}{6}$ yard to decorate a present. How many yards does he have left?

$$\frac{5}{6} - \frac{2}{6} = y$$

(number model)

Answer: ___$\frac{3}{6}$___ yard

SRB
178-180,
186

2 Solve.

a. 4 7 4
   ×   4
   ———
   1,896

b. 4 7 4
   ×   8
   ———
   3,792

SRB
100,
102, 104

3 Look at each number sentence. Choose True or False.

a. $16 - (3 + 5) = 18$

    ○ True   ● False

b. $(4 + 2) * 5 = 30$

    ● True   ○ False

c. $100 ÷ (25 + 25) + 5 = 7$

    ● True   ○ False

d. $\{(40 - 4) ÷ 6\} + 8 = 14$

    ● True   ○ False

SRB
42-43

4 Use the fractions listed. Fill in the blanks to make the number sentences true.

$\frac{1}{10}$ $1\frac{1}{4}$ $\frac{1}{8}$

Sample answers:

a. $\frac{1}{2} + $ ___$1\frac{1}{4}$___ $ < 2$

b. $2\frac{1}{3} - $ ___$\frac{1}{10}$___ $ > 2$

c. $\frac{4}{5} + $ ___$\frac{1}{8}$___ $ < 1$

SRB
181-184

5 **Writing/Reasoning** Explain how you know the number sentence you wrote for Problem 4b is true.

Sample answer: I know that $\frac{1}{10}$ is less than $\frac{1}{3}$, so subtracting $\frac{1}{10}$ from $2\frac{1}{3}$ would leave a difference that is more than 2.

SRB
10-11,
181-184

① 5.NF.2 ② 5.NBT.5 ③ 5.OA.1 ④ 5.NF.2

104   ⑤ 5.NF.2, SMP3, SMP6

Fraction-Of Problems

Work with a partner or a small group to solve the problems. You can use counters, drawings, or number lines to help you. Be prepared to explain how you solved the problems.

SRB
195

1. There are 56 beads on a necklace. $\frac{1}{4}$ of the beads are blue. How many beads are blue?
   See *Teacher's Lesson Guide*, pages 301 and 302 for sample representations for Problems 1–3.

   ___14___ beads

2. Jenna had 45 yards of yarn. She used $\frac{1}{5}$ of it for a knitting project. How much yarn did she use?

   ___9___ yards

3. Morris has a rectangular garden with an area of 60 square feet. $\frac{1}{10}$ of the garden is planted with bean plants. How many square feet are planted with bean plants?

   ___6___ square feet

Try This

4. The length of my living room is 24 feet. The width of my living room is $\frac{1}{2}$ the length. What is the area of my living room?

   ___288___ square feet

5.NF.4, 5.NF.4a, 5.NF.6, SMP2

Using Models to Estimate Volumes

A mathematical model of each real-world object is given below.

For each object, use the mathematical model to estimate its volume. Be sure to include a unit when you write the volume.

SRB
233-234

Then write one or more number sentences to show how you found the volume.

1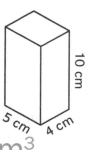

10 cm

5 cm   4 cm

Volume: __200 cm³__

__5 * 4 * 10 = 200__
(number sentence)

2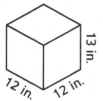

13 in.

12 in.   12 in.

Volume: __1,872 in.³__

__12 * 12 * 13 = 1,872__
(number sentence)

3

12 in.   1 in.

3 in.

Volume: __36 in.³__

__12 * 3 * 1 = 36__
(number sentence)

4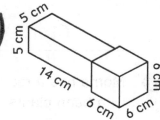

5 cm
5 cm

14 cm

6 cm

6 cm   6 cm

Volume: __566 cm³__

__6 * 6 * 6 = 216; 14 * 5 * 5 =__
__350; 216 + 350 = 566__
(number sentences)

5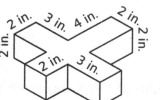

2 in.   3 in. 4 in. 2 in.
2 in.   2 in.
2 in.   2 in.   3 in.

Volume: __56 in.³__   Sample answer:

__2 * 2 * 2 = 8; 8 * 2 * 2 = 32;__
__2 * 2 * 4 = 16; 8 + 32 + 16__
__= 56__   (number sentences)

6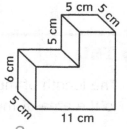

5 cm   5 cm

5 cm

6 cm

5 cm

11 cm

Volume: __455 cm³__

Sample answer: __11 * 6 * 5 =__
__330; 5 * 5 * 5 = 125; 330 +__
__125 = 455__ (number sentences)

106   5.MD.3, 5.MD.5, 5.MD.5a, 5.MD.5b, 5.MD.5c, SMP4

Math Boxes

1 Estimate and solve.

<u>Answers vary.</u>
(estimate)

$43\overline{)1,298}$

$1,298 \div 43 \rightarrow$ <u>30 R8</u>

SRB
84,
109-110

2 Use estimation strategies to determine whether the number sentences are true or false.

a. $\frac{2}{3} - \frac{1}{4} < 1\frac{1}{2}$ ⬤ True ◯ False

b. $\frac{1}{7} + \frac{2}{3} < 1$ ⬤ True ◯ False

c. $\frac{1}{3} + \frac{3}{4} < \frac{1}{2}$ ◯ True ⬤ False

d. $\frac{5}{4} - \frac{1}{10} > 1\frac{1}{2}$ ◯ True ⬤ False

SRB
181-182

3 Solve. Use fraction circle pieces to help you.

a. $\frac{3}{4} + \frac{1}{8} =$ <u>$\frac{7}{8}$</u>

b. $\frac{1}{2} + \frac{5}{8} =$ <u>$\frac{9}{8}$</u>, or $1\frac{1}{8}$

c. $\frac{2}{5} + \frac{3}{10} =$ <u>$\frac{7}{10}$</u>

d. $\frac{1}{2} + \frac{2}{5} =$ <u>$\frac{9}{10}$</u>

SRB
166, 189

4 Julio bought 4 pounds of carrot sticks and 3 pounds of celery sticks for the school potluck. He had 6 bowls for serving vegetables. He wants to put an equal amount of carrots and celery in each bowl.

Write an expression to show how many pounds of vegetables would be in each bowl.

<u>(4 + 3) / 6</u>

SRB
42, 44

5 Dixie wants to buy new stuffing for her couch. The drawing below is a model of her couch. Use the model to estimate the volume of her couch.

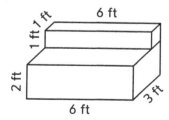

6 ft, 1 ft, 1 ft, 2 ft, 6 ft, 3 ft

Volume = <u>42</u> cubic feet

SRB
233-234

6 Each of the 17 gorillas at the nature preserve eats 45 pounds of food a day. Write an expression that models the amount of food all of the gorillas eat in a day. Then evaluate the expression to find how much food they eat.

Expression: <u>17 × 45</u>

Answer: <u>765</u> pounds

SRB
38, 44,
100-104

① 5.NBT.6 ② 5.NF.2 ③ 5.NF.1 ④ 5.OA.1, 5.OA.2 ⑤ 5.MD.5, 5.MD.5b, 5.MD.5c ⑥ 5.OA.2, 5.NBT.5

Math Boxes

More Fraction-Of Problems

Solve each problem. You can use drawings to help. Be sure to check that your answers make sense. **Sample drawings are given.**

SRB
195-196

1 What is $\frac{1}{2}$ of 5?

Answer: $\frac{5}{2}$, or $2\frac{1}{2}$

2 What is $\frac{1}{5}$ of 12?

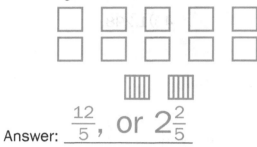

Answer: $\frac{12}{5}$, or $2\frac{2}{5}$

3 What is $\frac{1}{4}$ of 2?

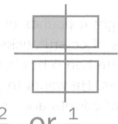

Answer: $\frac{2}{4}$, or $\frac{1}{2}$

4 What is $\frac{1}{8}$ of 6?

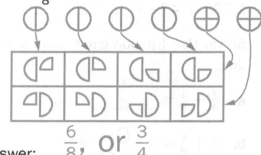

Answer: $\frac{6}{8}$, or $\frac{3}{4}$

5 Fredrick went to a farmer's market and bought 8 quarts of strawberries. He wants to keep $\frac{1}{3}$ of the strawberries and give away the rest. How many quarts of strawberries will he keep?

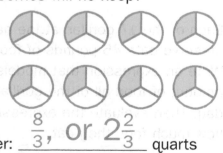

Answer: $\frac{8}{3}$, or $2\frac{2}{3}$ quarts

6 Shelby has a 4-pound bag of mixed nuts. She wants to put $\frac{1}{6}$ of the nuts at each of the 6 snack tables at a fundraiser. How many pounds of nuts should she put on each table?

Answer: $\frac{4}{6}$, or $\frac{2}{3}$ pound

7 Explain how you checked that your answer to Problem 6 made sense.

Sample answer: If $\frac{2}{3}$ is $\frac{1}{6}$ of 4, then $\frac{2}{3} + \frac{2}{3} + \frac{2}{3} + \frac{2}{3} + \frac{2}{3} + \frac{2}{3}$ should be $\frac{6}{6}$ of 4, or equal to 4. When I add it up, I get $\frac{12}{3}$, which is the same as 4. So my answer makes sense.

5.NF.4, 5.NF.4a, 5.NF.6, SMP1, SMP2

Math Boxes

1 Emma and her friends hiked $1\frac{1}{3}$ miles on Saturday. The next day they hiked $\frac{2}{3}$ mile. How many miles did they hike in all?

$$1\frac{1}{3} + \frac{2}{3} = p$$

(number model)

Answer: $1\frac{3}{3}$, or 2 miles

SRB
178-180,
186-187

2 Solve.

a.
```
    3 7 5
  ×     3
  ───────
  1,125
```

b.
```
    3 7 5
  ×     9
  ───────
  3,375
```

SRB
100,
102, 104

3 Solve.

a. $(28 / 7) * 3 = $ __12__

b. $\{(14 / 7) + (12 / 6)\} * 5 = $ __20__

c. $(3 * 10^2) + (5 * 2) = $ __310__

d. $32 + \{(8 * 2) / (2 + 2)\} = $ __36__

SRB
42-43

4 Use the fractions listed. Fill in the blanks to make each number sentence true.

$\frac{1}{3}$ $\frac{1}{6}$ $\frac{2}{7}$ $2\frac{3}{4}$ $\frac{1}{8}$ $1\frac{2}{3}$

Sample answers:

a. $\dfrac{\frac{1}{6}}{} + \dfrac{\frac{2}{7}}{} < \frac{1}{2}$

b. $\dfrac{1\frac{2}{3}}{} - \dfrac{\frac{1}{8}}{} > 1\frac{1}{2}$

c. $\dfrac{2\frac{3}{4}}{} + \dfrac{\frac{1}{3}}{} > 2\frac{1}{2}$

SRB
181-184

5 **Writing/Reasoning** Write a number story that could be modeled by the number sentence in Problem 3a.

Sample answer: There were 28 kids in a class. They divided into 7 groups for snack time. Each group got 3 packages of crackers. How many packages of crackers did the class need for snack?

SRB
42-44

① 5.NF.2 ② 5.NBT.5 ③ 5.OA.1 ④ 5.NF.2
⑤ 5.OA.2, SMP2

109

Math Boxes

1 Which shows expanded form for the number 942,462?
Choose the best answer.

◯ 94 × 10,000 + 24 × 100 + 62 × 10

⬤ 9 [100,000s] + 4 [10,000s] + 2 [1,000s] + 4 [100s] + 6 [10s] + 2 [1s]

◯ 9 [10,000s] + 4 [1,000s] + 2 [100s] + 4 [10s] + 6 [1s] + 2 [0s]

SRB
70

2 **a.** Round 318 to the nearest ten.
320

b. Round 4,135 to the nearest hundred.
4,100

c. Round 23,891 to the nearest thousand.
24,000

SRB
79-82

3 Write the money amounts in dollars-and-cents notation.

ten dollars and fifteen cents
$10.15

six dollars and eight cents
$6.08

Circle the amount that is greater.

$217.93   （$217.95）

SRB
121-123

4 Write the number that each point represents on the number line.

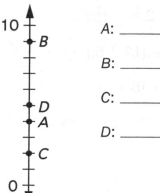

A: 4

B: 9

C: 2

D: 5

SRB
275

5 Write each fraction as a decimal.

a. $\frac{32}{100}$ 0.32

b. $\frac{9}{10}$ 0.9

c. $\frac{10}{100}$ 0.10

SRB
116

6 What is the value of the bold digit?

a. $32.4**2**   2 cents

b. $1**1**6.26   10 dollars

c. $0.8**6**   80 cents

SRB
118-119

① 5.NBT.1, 5.NBT.3, 5.NBT.3a ② 5.NBT.4 ③ 5.NBT.3,
5.NBT.3a, 5.NBT.3b ④ 5.G.2 ⑤ 5.NBT.3, 5.NBT.3a

⑥ 5.NBT.3, 5.NBT.3a, 5.NBT.3b

Math Boxes

1 Place the fractions on the number line.

$$\frac{7}{2} \qquad \frac{3}{2} \qquad \frac{5}{2}$$

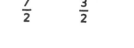

0   1 $\frac{3}{2}$ 2 $\frac{5}{2}$ 3 $\frac{7}{2}$ 4

Write $\frac{7}{2}$ as a mixed number. $3\frac{1}{2}$

SRB
159-160,
171

2 Make an estimate. Then use U.S. traditional multiplication to solve.

Answers vary.

(estimate)

```
      1
      1
    3 0 2
*     8 9
  2 7 1 8
+ 2 4 1 6 0
  2 6,8 7 8
```

SRB
83, 103

3 Beatrice has a page of 8 math problems to complete over the weekend. She solved 3 problems on Saturday morning and 2 on Saturday afternoon.

Write a number sentence with fractions that describes how much of the page Beatrice has completed.

$$\frac{3}{8} + \frac{2}{8} = p$$

What fraction of the page did Beatrice complete on Saturday?

$\frac{5}{8}$ of the page

SRB
178-180,
186

4 Solve.

a. $\frac{1}{2}$ of 6 = 3

b. $\frac{1}{2}$ of 8 = 4

c. $\frac{1}{3}$ of 12 = 4

d. $\frac{1}{4}$ of 20 = 5

SRB
195

5 **Writing/Reasoning** Choose a fraction from Problem 1. Write a division number story that has that fraction as the answer.

Sample answer: For $\frac{5}{2}$: Joe and his sister are allowed to have the television on for 5 hours a week. If they split the time evenly between them, how many hours of television do they each get to watch?

SRB
163-164

① 5.NF.3 ② 5.NBT.5 ③ 5.NF.2 ④ 5.NF.4, 5.NF.4a
⑤ 5.NF.3, SMP1, SMP2

Reading and Writing Decimals

Lesson 4-1

DATE   TIME

Ones 1s 1s $\frac{}{}$	.	Tenths 0.1s $\frac{1}{10}$s	Hundredths 0.01s $\frac{1}{100}$s	Thousandths 0.001s $\frac{1}{1,000}$s
	.			
0	.	3	4	5
1	.	0	7	8
5	.	1	6	
0	.	8		
2	.	3	0	7
7	.	0	9	0
4	.	8		
0	.	4	8	
0	.	0	4	8
6	.	4	0	8
	.			

5.NBT.1, 5.NBT.3, 5.NBT.3a, SMP2, SMP7

Write the following decimals in words. Use the place-value chart on journal page 112 to help you.

SRB
116-119

1 0.67 __sixty-seven hundredths__

2 3.8 __three and eight tenths__

3 3.622 __three and six hundred twenty-two thousandths__

4 0.804 __eight hundred four thousandths__

Write each decimal using numerals. Record them on the place-value chart on page 112. Then write the value of 4 in each decimal.

5 a. four and eight tenths __4.8__   b. 4 is worth __4 ones, or 4__.

6 a. forty-eight hundredths __0.48__   b. 4 is worth __4 tenths, or 0.4__.

7 a. forty-eight thousandths __0.048__   b. 4 is worth __4 hundredths, or 0.04__.

8 a. six and four hundred eight thousandths __6.408__

   b. 4 is worth __4 tenths, or 0.4__.

Rewrite each decimal as a fraction.

9 0.6 __$\frac{6}{10}$__   10 0.03 __$\frac{3}{100}$__   11 0.008 __$\frac{8}{1,000}$__

Rewrite each fraction as a decimal.

12 $\frac{2}{10}$ __0.2__   13 $\frac{65}{1,000}$ __0.065__   14 $\frac{402}{1,000}$ __0.402__

15 Use the clues to write the mystery number.

Write 5 in the tenths place.

Write 6 in the ones place.

Write 2 in the thousandths place.

Write 1 in the hundredths place.

__6__.__5__ __1__ __2__

16 Make the following changes to the number 7.849:

Make the 7 worth $\frac{1}{10}$ as much.

Make the 8 worth 10 times as much.

Make the 4 worth $\frac{1}{10}$ as much.

Make the 9 worth 10 times as much.

__8__.__7__ __9__ __4__

Representing Decimals

Math Message

SRB
116-118, 120

(1) Write 0.43 in words. <u>forty-three hundredths</u>

(2) Write 0.43 in the place-value chart below.

Ones	.	Tenths	Hundredths	Thousandths
0	.	4	3	

(3) The grid below represents 1. Shade the grid to show 0.43.

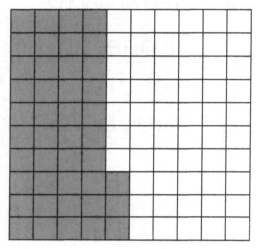

For Problems 4–6, use words, fractions, equivalent decimals, or other representations to write at least three names for each decimal in the name-collection box. Then shade the grid to show the decimal. Sample names given.

(4)

0.8
0.80
0.800
eight tenths
$\frac{8}{10}$
$\frac{80}{100}$

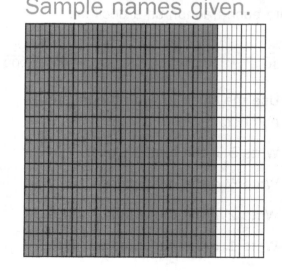

5.NBT.1, 5.NBT.3, 5.NBT.3a, SMP2

⑤

0.620
0.62
sixty-two hundredths
six hundred twenty thousandths
$\frac{62}{100}$
$\frac{620}{1,000}$

Sample names given.

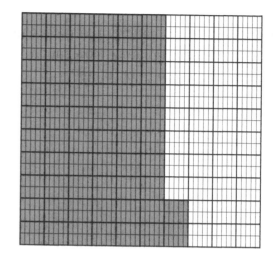

⑥

0.418
$\frac{418}{1,000}$
four hundred eighteen thousandths
418 thousandths
0.4180
$\frac{4,180}{10,000}$

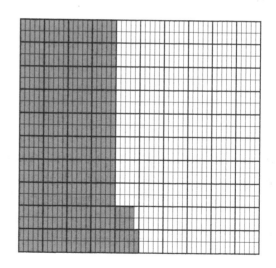

5.NBT.1, 5.NBT.3, 5.NBT.3a, SMP2

115

Fraction Number Stories

For each story, write a number model with a letter for the unknown. Then solve the story. You can draw pictures or use fraction circle pieces to help.

SRB
163-164,
178-180

---

1 A chef divides 5 heads of lettuce equally among 12 salad plates. How much lettuce will be on each plate?

Number model: $5 \div 12 = l$

Answer: $\dfrac{5}{12}$ head of lettuce

2 Alma has a 10-foot roll of wrapping paper. She cuts off $2\frac{3}{4}$ feet of paper to wrap a gift. How much paper is left on the roll?

Number model: $10 - 2\frac{3}{4} = f$

Answer: $7\frac{1}{4}$ feet of paper

---

3 Vernon mixed $2\frac{1}{3}$ cups of water with $2\frac{1}{3}$ cups of white vinegar to make a cleaning solution. How much cleaning solution did he make?

Number model: $2\frac{1}{3} + 2\frac{1}{3} = c$

Answer: $4\frac{2}{3}$ cups

4 In a relay race one runner ran $4\frac{3}{10}$ miles. The next runner ran $4\frac{9}{10}$ miles. What is the total distance they ran?

Number model: $4\frac{3}{10} + 4\frac{9}{10} = m$

Answer: $9\frac{2}{10}$ miles

---

5 A guinea pig weighs $1\frac{7}{8}$ pounds. A rabbit weighs $3\frac{3}{8}$ pounds. How much more does the rabbit weigh than the guinea pig?

Number model: $3\frac{3}{8} - 1\frac{7}{8} = w$

Answer: $1\frac{4}{8}$, or $1\frac{1}{2}$ pounds

6 A bicycle relay race is 24 miles long. Nikita's team has 7 members who will each ride the same distance in the race. How far will each team member ride?

Number model: $24 \div 7 = b$

Answer: $3\frac{3}{7}$ miles

---

116   5.NF.2, 5.NF.3, SMP1, SMP4

Math Boxes

1 Write each number in expanded form.
Sample answers given.

a. 21,756,834 $2 \times 10,000,000 +$
$1 \times 1,000,000 + 7 \times 100,000 +$
$5 \times 10,000 + 6 \times 1,000 +$
$8 \times 100 + 3 \times 10 + 4 \times 1$

b. 311,019
$3 [100,000s] + 1 [10,000s] +$
$1 [1,000s] + 1 [10s] + 9 [1s]$

SRB
70

2 Estimate. Then use partial-quotients division to solve.

$2,731 \div 31 = ?$

_____Answers vary._____
(estimate)

$2,731 \div 31 \rightarrow$ __88 R3__

SRB
84,
109-110

3 Find the volume of the rectangular prism.
Use the formula $V = l * w * h$.

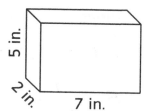

5 in.

2 in.   7 in.

$V =$ __$7 \times 2 * 5$__
(number model)

$V =$ __70__ in.³

SRB
233

4 **a.** Rewrite the height of each student in *inches only*.

Juan: 5 feet, 4 inches = __64 inches__

Marilu: 4 feet, 11 inches = __59 inches__

b. How many inches taller than Marilu is Juan?

__5__ inches

SRB
215-217,
219, 328

5 Write >, <, or =.

a. $\frac{2}{3} - \frac{1}{2}$ __<__ $\frac{1}{2}$

b. $\frac{3}{4} + \frac{3}{8}$ __>__ 1

c. $\frac{3}{4} - \frac{1}{2}$ __=__ $\frac{1}{4}$

SRB
181-182

6 The swimming pool is open 8 hours a day. The pool manager must divide pool time equally among 5 groups: camp, swim team, swim lessons, family swim, and open swim. How much time should she allow for each group?

__$\frac{8}{5}$, or $1\frac{3}{5}$__ hours

SRB
163-164

① 5.NBT.1 ② 5.NBT.6 ③ 5.MD.5, 5.MD.5a, 5.MD.5b
④ 5.MD.1 ⑤ 5.NF.2 ⑥ 5.NF.3

Writing Decimals in Expanded Form

Math Message

Shade the grid to represent 0.3. Add shading to the grid in another color so that the grid shows 0.31 in all. Then use a third color to add shading so that the grid shows 0.312.

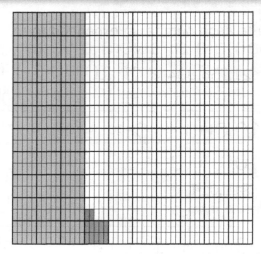

Use different versions of expanded form to complete the table below.

Standard Notation	Versions of Expanded Form		
	Sum of Decimals in Standard Notation	Sum of Multiplication Expressions (Decimals)	Sum of Multiplication Expressions (Fractions)
Example: 0.568	0.5 + 0.06 + 0.008	(5 * 0.1) + (6 * 0.01) + (8 * 0.001)	$5 * \frac{1}{10} + 6 * \frac{1}{100} + 8 * \frac{1}{1,000}$
2.473	2 + 0.4 + 0.07 + 0.003	(2 * 1) + (4 * 0.1) + (7 * 0.01) + (3 * 0.001)	$(2 * 1) + \left(4 * \frac{1}{10}\right) + \left(7 * \frac{1}{100}\right) + \left(3 * \frac{1}{1,000}\right)$
0.094	0.0 + 0.09 + 0.004	(0 * 0.1) + (9 * 0.01) + (4 * 0.001)	$\left(0 * \frac{1}{10}\right) + \left(9 * \frac{1}{100}\right) + \left(4 * \frac{1}{1,000}\right)$
7.752	7 + 0.7 + 0.05 + 0.002	(7 * 1) + (7 * 0.1) + (5 * 0.01) + (2 * 0.001)	$(7 * 1) + \left(7 * \frac{1}{10}\right) + \left(5 * \frac{1}{100}\right) + \left(2 * \frac{1}{1,000}\right)$
0.637	0.6 + 0.03 + 0.007	(6 * 0.1) + (3 * 0.01) + (7 * 0.001)	$\left(6 * \frac{1}{10}\right) + \left(3 * \frac{1}{100}\right) + \left(7 * \frac{1}{1,000}\right)$

Representing Decimals in Expanded Form

Use three number cards to create a decimal on your decimal place-value mat. Record the decimal you created in one of the boxes below. Write the decimal in expanded form. Then shade a thousandths grid using a different color to show the value of each digit. Repeat to complete all four boxes. Answers vary.

Decimal: 0. _____ _____ _____

Expanded form: _____

_____

Decimal: 0. _____ _____ _____

Expanded form: _____

_____

Decimal: 0. _____ _____ _____

Expanded form: _____

_____

Decimal: 0. _____ _____ _____

Expanded form: _____

_____

5.NBT.1, 5.NBT.3, 5.NBT.3a, SMP2

119

Math Boxes

Math Boxes

1 Label each mark on the number line with the appropriate fraction below.

$$\frac{4}{3} \qquad \frac{3}{2} \qquad \frac{7}{4}$$

1     $\frac{4}{3}$ $\frac{3}{2}$ $\frac{7}{4}$ 2

Write $\frac{4}{3}$ as a mixed number. ___$1\frac{1}{3}$___

2 Make an estimate. Then use U.S. traditional multiplication to fill in the missing numbers.

_____ Answers vary. _____
(estimate)

```
              2   1
              1   1
              8   7   5
  *           1   3   2
  _____
          1   7   5   0
      2   6   2   5   0
  +   8   7   5   0   0
  _____
  1   1   5 , 5   0   0
```

3 Dawn wants to plant vegetables and herbs in her garden. If she plants herbs in $\frac{3}{8}$ of the garden, how much of the garden is left over for vegetables?

Sample answer: $\frac{8}{8} - \frac{3}{8} = g$
(number model)

Answer: ___$\frac{5}{8}$___ garden

4 What is:

a. $\frac{1}{3}$ of 24? ___8___

b. $\frac{1}{4}$ of 24? ___6___

c. $\frac{1}{6}$ of 24? ___4___

5 **Writing/Reasoning** Explain how you solved Problem 4c.

Sample answer: I drew 6 circles and put one X in each circle until I had counted out 24 Xs. There were 4 Xs in each circle, so I knew that $\frac{1}{6}$ of 24 is 4.

① 5.NF.3 ② 5.NBT.5 ③ 5.NF.2 ④ 5.NF.4, 5.NF.4a

⑤ 5.NF.4, 5.NF.4a, SMP6

Interpreting Remainders

Solve each number story. Show your work. Explain what you decided to do with the remainder.

SRB 109, 113-114

1. Bre earned 189 tickets playing different games at the fair. If each prize costs 15 tickets, how many prizes can Bre get?

   Number model: $189 \div 15 = p$

   Quotient: ___12___ Remainder: ___9___

   Answer: Bre can get ___12___ prizes.

   Circle what you did with the remainder.

   (Ignored it)   Reported it   Rounded the
                       as a fraction   quotient up

   Why? Sample answer: Bre doesn't have enough tickets left over for another prize.

   Sample work:

   $$15\overline{)189}$$
   $$-\ 150 \quad | \quad 10$$
   $$\overline{39}$$
   $$-\ 30 \quad | \quad 2$$
   $$\overline{9} \quad 12$$

2. Elisbeth is 58 inches tall. What is her height in feet?

   Remember: 1 foot = 12 inches

   Number model: $58 \div 12 = h$

   Quotient: ___4___ Remainder: ___10___

   Answer: Elisbeth is ___$4\frac{10}{12}$___ feet tall.

   Circle what you did with the remainder.

   Ignored it   (Reported it   Rounded the
                      as a fraction)   quotient up

   Why? Sample answer: Rounding up or ignoring the remainder would change Elisbeth's height. We want to know the number of feet that is exactly equal to her height in inches.

   Sample work:

   $$12\overline{)58}$$
   $$-\ 48 \quad | \quad 4$$
   $$\overline{10} \quad 4$$
   $$10 \div 12 = \frac{10}{12}$$

5.NBT.6, 5.NF.3, SMP6

121

Math Boxes

① Write the following number in expanded form.

3,768,412,000 Sample answer:

3 × 1,000,000,000 +
7 × 100,000,000 +
6 × 10,000,000 +
8 × 1,000,000 +
4 × 100,000 + 1 ×
10,000 + 2 × 1,000
+ 0 × 100 +
0 × 10 + 0 × 1

SRB
70

② Estimate. Then use partial-quotients division to solve.

8,096 ÷ 21 = ?

Answers vary.
_____
(estimate)

8,096 ÷ 21 → 385 R11

SRB
84,
109-110

③ Find the volume of the rectangular prism. Use the formula $V = B * h$.

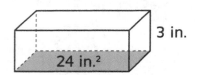

3 in.

24 in.²

$V = $ 72 cubic inches

SRB
233

④ **a.** Rewrite each baby's birth weight in ounces (oz) only.

Remember: 16 oz = 1 lb

Sadie: 5 lb, 11 oz = 91 oz

Kevin: 8 lb, 9 oz = 137 oz

b. How many more ounces did Kevin weigh at birth than Sadie?

46 oz

SRB
215-217

⑤ Write >, <, or =.

a. $1\frac{1}{3} - \frac{2}{3}$ < 1

b. $\frac{3}{4} - \frac{1}{8}$ > $\frac{1}{2}$

c. $\frac{4}{8} + \frac{6}{12}$ = 1

SRB
181-182

⑥ Matt has a job as a dog walker. He has 4 hours to walk 18 dogs. Which expressions represent the amount of time that should be spent walking each dog? Select all that apply.

☐ $\frac{18}{4}$ hours   ☐ (18 ÷ 4) hours

☑ $\frac{4}{18}$ hour   ☑ (4 ÷ 18) hour

SRB
163-164

① 5.NBT.1 ② 5.NBT.6 ③ 5.MD.5, 5.MD.5b ④ 5.MD.1

Identifying the Closer Number

Math Message

SRB
124-125

1   Shade the grid at the right to show 0.28.

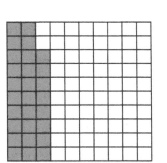

2   Is 0.28 closer to 0.2 or 0.3? Use the grid to help you decide. Be ready to explain your reasoning.

0.28 is closer to ___0.3___.

---

3   Label the number line below to show whether 0.28 is closer to 0.2 or 0.3.

4   Shade the grid at the right to show 0.619. Use the grid to help you solve Problem 5.

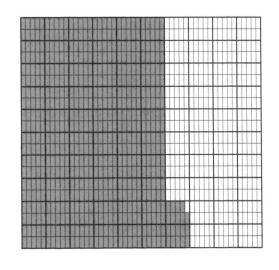

5   **a.**   Between which two hundredths is 3.619?

3.619 is between ___3.61___ and ___3.62___.

     b.   What number is exactly halfway between the numbers you wrote in Problem 5a?

___3.615___

6   Round 3.619 to the nearest hundredth. Use your answers from Problems 4 and 5 and the number line below to help you.

___3.62___

Round up.

5.NBT.1, 5.NBT.3, 5.NBT.3a, 5.NBT.3b, 5.NBT.4, SMP2   123

Rounding Decimals

Use the number lines to round each number. Be sure to label the tick marks on each number line.

1 Round 3.6 to the nearest whole number.

SRB
125

3   3.5 3.6   4

Rounded number: _____4_____ Did you round up or down? ___Up___

2 Round 2.73 to the nearest tenth.

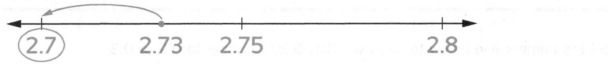

2.7   2.73   2.75   2.8

Rounded number: ___2.7___ Did you round up or down? ___Down___

3 Round 2.73 to the nearest whole number.

2   2.5   2.73   3

Rounded number: _____3_____ Did you round up or down? ___Up___

4 Round 4.254 to the nearest hundredth.

4.25   4.255   4.26
              4.254

Rounded number: ___4.25___ Did you round up or down? ___Down___

5 Round 4.254 to the nearest tenth.

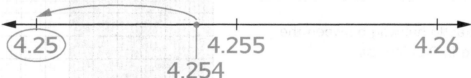

4.2   4.25   4.3
              4.254

Rounded number: ___4.3___ Did you round up or down? ___Up___

6 Round 4.254 to the nearest whole number.

4   4.254   4.5   5

Rounded number: _____4_____ Did you round up or down? ___Down___

5.NBT.1, 5.NBT.3, 5.NBT.3a, 5.NBT.3b, 5.NBT.4, SMP2

Rounding Decimals in Real-World Contexts

Read about different real-world situations below and round the decimals as directed. Use a number line or grids to help you, if needed.

SRB
124-127

1 At the district track meet each running event is timed to the nearest thousandth of a second using an electronic timer. However, the district's track rules require times to be reported with only 2 decimal places. Round each time to the nearest hundredth of a second.

Note: sec = second(s)

Electronic Timer	Reported Time	Electronic Timer	Reported Time
a. 10.752 sec	10.75 sec	**b.** 55.738 sec	55.74 sec
c. 16.815 sec	16.82 sec	**d.** 43.505 sec	43.51 sec
e. 20.098 sec	20.10 sec	**f.** 52.996 sec	53.00 sec

Explain how you rounded 20.098 to the nearest hundredth. Sample answer: I knew that 8 thousandths was more than halfway to the next hundredth, so I decided to round up. Rounding 9 hundredths up meant having 10 hundredths, which is equal to 1 tenth, so I wrote 20.10.

2 Supermarkets often show unit prices for items. This helps customers compare prices to find the best deal. A unit price is found by dividing the price of an item (in cents or dollars and cents) by the quantity of the item (often in ounces or pounds). When the quotient has more decimal places than are needed, some stores round to the nearest tenth of a cent.

Example: A 16 oz container of yogurt costs $3.81.

• $3.81 * 100 cents per dollar = 381¢

• 381¢ ÷ 16 oz = 23.8125¢ per ounce

• 23.8125¢ is rounded down to 23.8¢ per ounce

Round each unit price to the nearest tenth of a cent.

a. 28.374¢ ___28.4___ ¢

b. 19.756¢ ___19.8___ ¢

c. 16.916¢ ___16.9___ ¢

d. 20.641¢ ___20.6___ ¢

e. 18.459¢ ___18.5___ ¢

f. 21.966¢ ___22.0___ ¢

5.NBT.1, 5.NBT.4

Math Boxes

1 Keegan practices karate for 60 minutes every weekday. If the summer has 47 weekdays, how many minutes will he have practiced by the end of summer?

$$60 \times 47 = t$$
(number model)

$$60 \times 50 = 3,000$$
(estimate)

Answer: _____ 2,820 minutes

SRB 44, 83, 100-104

2 Shade the grid to represent the decimal 0.8.

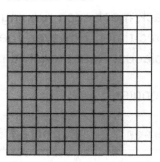

Write the decimal in words.
eight tenths, or eighty hundredths

SRB 117, 120

3 Circle the benchmark that is closest to each sum or difference.

a. $\frac{3}{8} + \frac{9}{10}$

0      $\frac{1}{2}$ 1 $\left(1\frac{1}{2}\right)$ 2

b. $1\frac{1}{6} - \frac{3}{5}$

0      $\left(\frac{1}{2}\right)$ 1 $1\frac{1}{2}$ 2

SRB 181-184

4 Olivia is buying bundles of wood for a campfire. The bundles of wood can't be split up. If each bundle costs \$3, how many bundles can she buy with \$10?

$$10 \div 3 = b$$
(number model)

Answer: _____ 3 _____ bundles of wood

SRB 44, 113

5 **Writing/Reasoning** Explain how you solved Problem 3b.
Sample answer: I saw that $1\frac{1}{6}$ is close to 1 and $\frac{3}{5}$ is close to $\frac{1}{2}$ so I thought about $1 - \frac{1}{2}$ and estimated the answer to be around $\frac{1}{2}$.

SRB 181-184

① 5.NBT.5 ② 5.NBT.3, 5.NBT.3a ③ 5.NF.2 ④ 5.NBT.6
⑤ 5.NF.2, SMP6

Locating Cities on a Map of Ireland

Bantry	B-1	Dublin	F-4	Lahinch	B-4	Omagh	E-7
Belfast	F-7	Dundalk	F-6	Larne	F-7	Tralee	B-2
Carlow	E-3	Galway	C-4	Limerick	C-3	Tuam	C-5
Castlebar	B-6	Gort	C-4	Mullingar	E-5	Westport	B-5
Derry	E-8	Kilkee	B-3	Navan	E-5	Wicklow	F-4

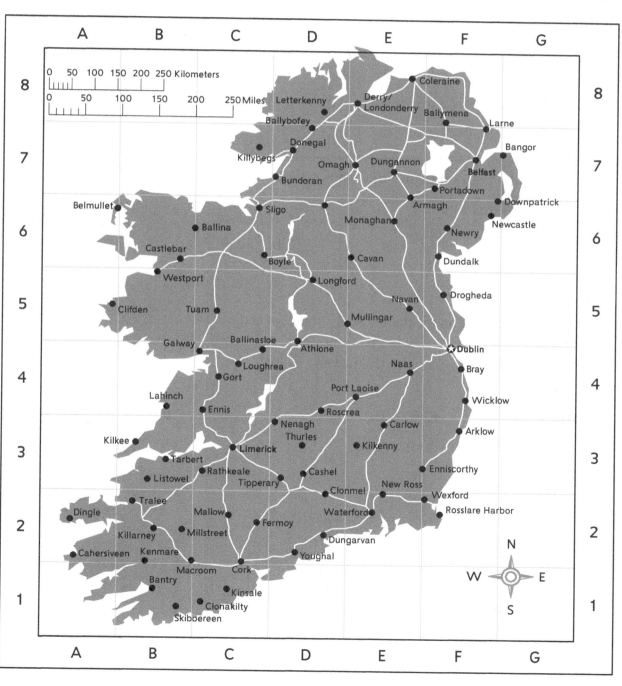

Plotting Points on a Coordinate Grid

1. Use the grid shown below as you follow directions from your teacher.

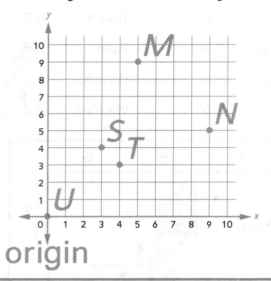

origin

For Problems 2–6, use the coordinate grid shown below.

2. Write the ordered pairs for:

    a. Point A ___(10, 13)___

    b. Point B ___(5, 13)___

    c. Point C ___(5, 9)___

    d. the origin ___(0, 0)___

3. Use a straightedge. Connect points A, B, and C in order.

4. On the same grid, plot and label the points listed below.

    D (6, 10)   E (9, 10)

    F (10, 9)   G (10, 6)

    H (9, 5)   J (6, 5)

    K (5, 6)

5. Use a straightedge. Connect the points in alphabetical order beginning with point C.

6. What image did you create by connecting the points?

    ___The number 5___

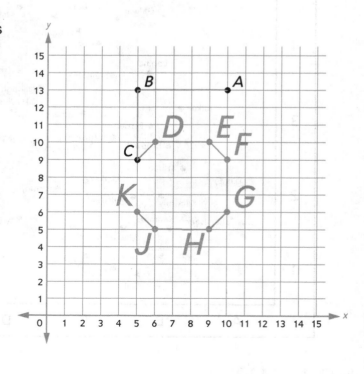

How Much Soil?

1 **a.** David helped his father build a large planter to grow flowers in front of their house. How much soil can the planter hold?

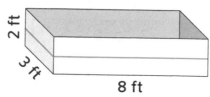

The planter can hold ____**48**____ cubic feet of soil.

Number model: _**8 * 3 * 2 = 48**_

b. David found 11 unopened bags of potting soil in the garage. Each bag contained 2 cubic feet of soil. David dumped all the bags of soil into the planter. Draw a line on the picture above to show about how much soil is in the planter. Explain how you figured it out.

Sample answer: 11 bags with 2 cubic feet in each bag is 22 cubic feet. 22 is just a little less than half of 48, so I drew the line a little less than halfway up the planter.

2 David's mother used scraps of wood to build the vegetable planter shown below for the back patio.

a. How much soil can the planter hold?

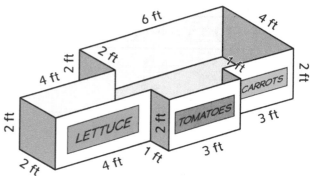

The vegetable planter can hold ____**70**____ cubic feet of soil.

b. Explain how you solved Part a. Sample answer: I divided the planter into 3 parts. To find how much the lettuce part could hold, I multiplied 4 * 2 * 2 to get 16. To find the volume of the tomato part, I multiplied 3 * 2 * 5 to get 30. The carrot part is 3 * 4 * 2 = 24. Then I added 16 + 30 + 24 to get 70 cubic feet.

c. If the planter doesn't fit on the patio, David's mom will remove the lettuce section. Then how much soil would be needed to fill the planter? ____**54**____ cubic feet

3 Talk to a partner about how you solved each problem. Compare your strategies.

5.MD.3, 5.MD.5, 5.MD.5a, 5.MD.5b, 5.MD.5c, SMP1, SMP4   129

Math Boxes

1 Sasha walked $\frac{3}{5}$ mile to school. After school, she walked another $\frac{1}{5}$ mile to ballet practice. From ballet, she walked $\frac{4}{5}$ mile home. How far did Sasha walk all together?

$$\frac{3}{5} + \frac{1}{5} + \frac{4}{5} = w$$

(number model)

Answer: $\underline{\frac{8}{5}, \text{ or } 1\frac{3}{5}, \text{ miles}}$

SRB 178-180, 186

2 Fill in the name-collection box with at least 3 names. Sample answers:

0.25
twenty-five hundredths
$(2 * 0.1) + (5 * 0.01)$
$\frac{25}{100}$
25 hundredths
$0.2 + 0.05$

SRB 116-118

3 Write 4.628 in expanded form.

Sample answer:

$$4 + 0.6 + 0.02 + 0.008$$

SRB 118

4 Circle the situation below that would have the answer: $\frac{8}{5}$ kilograms of krill.

A. 8 penguins eat 5 kilograms of krill. How much does each penguin eat?

B. 5 penguins eat 8 kilograms of krill. How much does each penguin eat?

SRB 163-164

5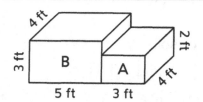

This figure models the steps to Shelby's porch. Which of the following are true? Fill in the circle next to all that apply.

⬤ **A.** The total volume of the steps is 84 ft³.

◯ **B.** The volume of step A is 28 ft³.

⬤ **C.** The volume of step B is 60 ft³.

SRB 233-234

6 Make an estimate. Then solve.

4,211 ÷ 68

Answers vary.

(estimate)

$4{,}211 \div 68 \rightarrow \underline{61 \text{ R}63}$

SRB 84, 109-110

① 5.NF.2 ② 5.NBT.3, 5.NBT.3a ③ 5.NBT.3, 5.NBT.3a
④ 5.NF.3 ⑤ 5.MD.5, 5.MD.5b, 5.MD.5c ⑥ 5.NBT.6

Town Map

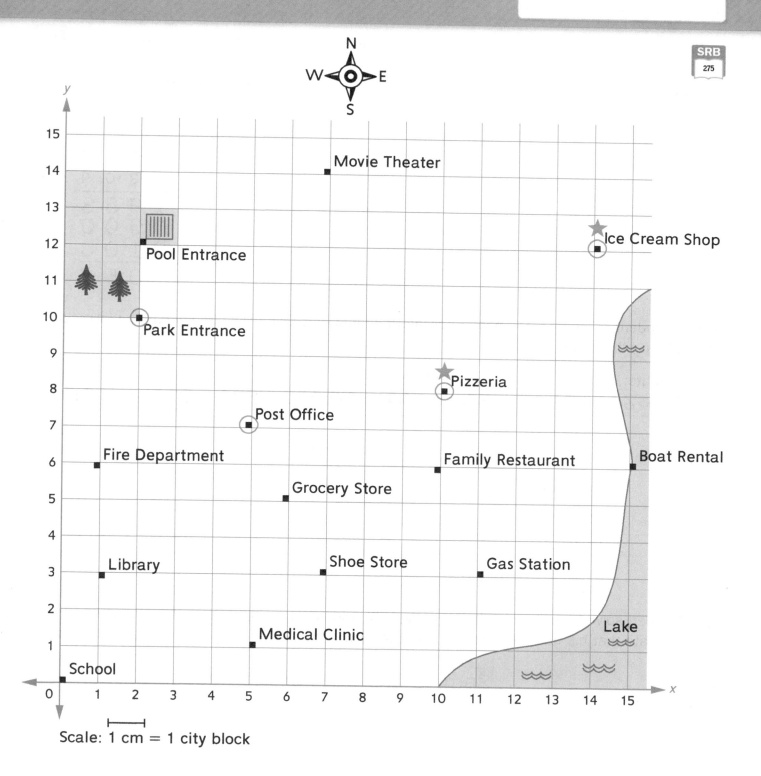

Scale: 1 cm = 1 city block

Practice with U.S. Traditional Multiplication

For Problems 1–3, make an estimate. Then solve using U.S. traditional multiplication.

SRB
83,
102-103

1 Estimate: Sample answer: 25 * 10 = 250

```
        2 5
    *   1 1
    ─────────
        2 5
    + 2 5 0
    ─────────
      2 7 5
```

2 Estimate: Sample answer: 400 * 30 = 12,000

```
        2 1
      4 3 2
    *     2 9
    ─────────
      3 8 8 8
    + 8 6 4 0
    ─────────
    1 2,5 2 8
```

3 Estimate: Sample answer: 200 * 300 = 60,000

```
            2
        1   6
        2 1 7
    *   3 0 9
    ─────────
      1 9 5 3
      0 0 0 0
    + 6 5 1 0 0
    ─────────
    6 7,0 5 3
```

Another student estimated and began solving Problems 4–6 using U.S. traditional multiplication. Finish solving the problems.

4 Estimate:
40 * 60 = 2,400

```
          4
        1
        3 7
    *   6 2
    ─────────
        7 4
    + 2 2 2 0
    ─────────
    2,2 9 4
```

5 Estimate:
500 * 100 = 50,000

```
        8   1
        7   1
        4 9 2
    *     9 8
    ─────────
      3 9 3 6
    + 4 4 2 8 0
    ─────────
    4 8,2 1 6
```

6 Estimate:
500 * 200 = 100,000

```
        5 1 1
    *   2 1 9
    ─────────
      4 5 9 9
      5 1 1 0
    + 1 0 2 2 0 0
    ─────────
    1 1 1,9 0 9
```

7 Stephen solved the problem below using U.S. traditional multiplication. He can tell from his estimate that his answer is wrong. Find Stephen's mistake and explain how he could fix it.

Estimate:
700 * 80 = 56,000

```
        5   1
        5   1
        6 7 2
    *     8 7
    ─────────
      4 7 0 4
    + 5 3 7 6
    ─────────
    1 0,0 8 0
```

Sample answer: For the second partial product, Stephen multiplied 672 by 8 instead of 80. He could fix it by thinking "80 times 2 equals 160" and writing 0 in the 1s column, 6 in the 10s column, and the 1 above the 10s, and keep going.

132   5.NBT.5, SMP3, SMP6

Math Boxes

1 Colette is filling 400 water balloons for a school picnic. Balloons come in bags of 25, 50, 150, and 250. Which set of bags will provide at least 400 balloons, with the fewest number of balloons left over?

Fill in the circle next to the best answer.

○ **A.** 15 bags of 25 balloons

● **B.** 8 bags of 50 balloons

○ **C.** 3 bags of 150 balloons

○ **D.** 2 bags of 250 balloons

SRB
97-104

2 Shade in the grid to represent the decimal 0.08.

Write the decimal in words.

__eight hundredths__

SRB
117, 120

3 Circle the benchmark that is closest to each sum or difference.

a. $\frac{5}{9} - \frac{3}{6}$

Ⓞ $\frac{1}{2}$ 1

b. $\frac{1}{2} + \frac{5}{6}$

0    $\frac{1}{2}$ ①

SRB
181-184

4 Lola has 17 photos to place in the school newsletter. She can fit 4 photos per page. How many pages long must the newsletter be to fit all the photos?

$$17 \div 4 = p$$

(number model)

Quotient: __4__ Remainder: __1__

Answer: __5__ pages

SRB
44, 113

5 **Writing/Reasoning** Explain how you decided what to do with the remainder for Problem 4.

Sample answer: I divided 17 by 4 and got 4 pages with 1 photo left over. Since all the photos have to be placed, I rounded up and added a fifth page to include the last photo. So my answer is 5 pages.

SRB
113

① 5.NBT.5 ② 5.NBT.3, 5.NBT.3a ③ 5.NF.2 ④ 5.NBT.6
⑤ 5.NBT.6, SMP6

Graphing Sailboats

1. Find the column labeled Original Sailboat in the table below. Plot the ordered pairs listed in the column on the grid titled Original Sailboat on the next page. Connect the points in the same order that you plot them. You should see the outline of a sailboat.

SRB
275

2. **a.** Fill in the missing coordinates for New Sailboat 1.

   b. How do you think New Sailboat 1 will be different from the Original Sailboat? Record a conjecture at the top of the column.

   c. Plot the ordered pairs for New Sailboat 1 on the next page. Connect the points in the same order that you plot them.

Original Sailboat	New Sailboat 1 Rule: Double each number of the original pair.	New Sailboat 2 Rule: Double the first number of the original pair. Leave the second number the same.	New Sailboat 3 Rule: Double the second number of the original pair. Leave the first number the same.
Conjecture:	Sample answer: It will look twice as big.	Sample answer: It will look twice as wide.	Sample answer: It will look twice as tall.
(8, 1)	(16, 2)	(16, 1)	(8, 2)
(5, 1)	(10, 2)	(10, 1)	(5, 2)
(5, 7)	(10, 14)	(10, 7)	(5, 14)
(1, 2)	(2 , 4)	(2 , 2)	(1 , 4)
(5, 1)	(10 , 2)	(10 , 1)	(5 , 2)
(0, 1)	(0 , 2)	(0 , 1)	(0 , 2)
(2, 0)	(4 , 0)	(4 , 0)	(2 , 0)
(7, 0)	(14 , 0)	(14 , 0)	(7 , 0)
(8, 1)	(16 , 2)	(16 , 1)	(8 , 2)

   d. Complete steps 2a–2c for New Sailboat 2.

   e. Complete steps 2a–2c for New Sailboat 3.

   Be sure to apply each rule to the coordinates from the **Original Sailboat**.

134   5.NF.5, 5.NF.5a, 5.G.1, 5.G.2, SMP3, SMP6

Graphing Sailboats (continued)

SRB
10-11,
275

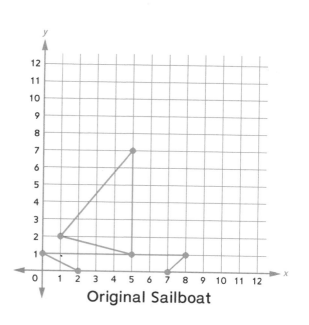

Original Sailboat

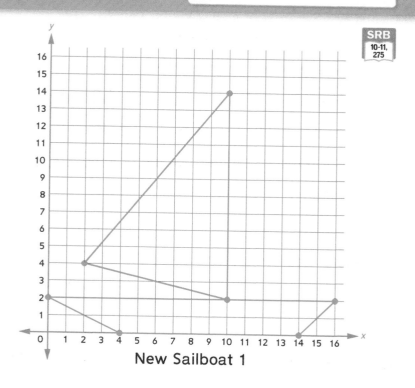

New Sailboat 1

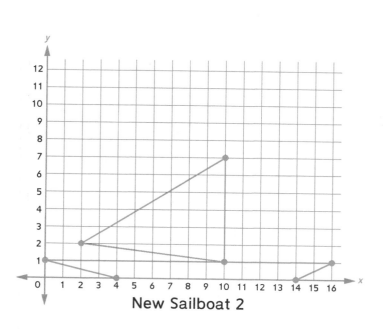

New Sailboat 2

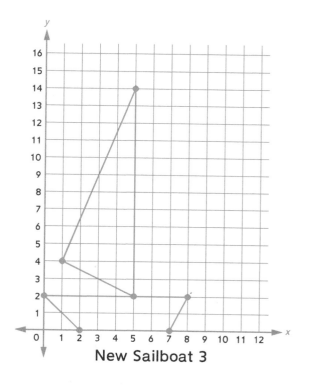

New Sailboat 3

5.NF.5, 5.NF.5a, 5.G.1, 5.G.2, SMP6

A New Sailboat Rule

1 Circle the rule for Sailboat 4 given to you by your teacher.

- Triple the first number of the ordered pair.

- Triple the second number of the ordered pair.

- Double the first number of the ordered pair; halve the second number.

- Halve the first number of the ordered pair; double the second number.

- Other: _____

_____

2 Make a conjecture about what New Sailboat 4 will look like.

_____ Answers vary. _____

_____

3 Create ordered pairs for New Sailboat 4 based on the rule. Write them in the table below.

4 Plot the new set of ordered pairs and connect the points in the order they were plotted.

5 Was your conjecture correct? Explain. _____ Answers vary. _____

_____

Answers vary.

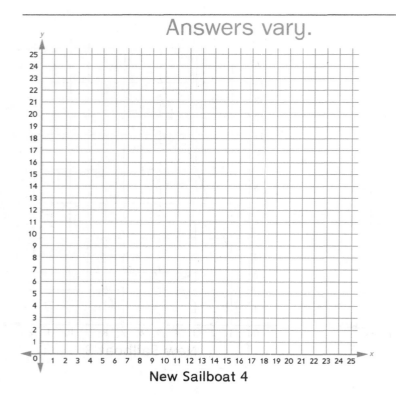

New Sailboat 4

Original Sailboat	New Sailboat 4
(8, 1)	(,)
(5, 1)	(,)
(5, 7)	(,)
(1, 2)	(,)
(5, 1)	(,)
(0, 1)	(,)
(2, 0)	(,)
(7, 0)	(,)
(8, 1)	(,)

Math Boxes

Math Boxes

1 Dakota had a kite with $25\frac{6}{12}$ feet of string. The kite got stuck in a tree. She lost $4\frac{3}{12}$ feet of string trying to free it. How much string does Dakota still have?

$$25\frac{6}{12} - 4\frac{3}{12} = k$$

(number model)

Answer: $21\frac{3}{12}$ ft of string

SRB
178-180

2 Complete the name-collection box.

Sample answers:

3.09
Three and nine hundredths
$3\frac{9}{100}$
$3 \times 1 + 0 \times 0.1 + 9 \times 0.01$
Three and ninety thousandths

SRB
116-118

3 Write the number 17.803 in expanded form.

Sample answer:
$1 \times 10 + 7 \times 1 + 8 \times 0.1 + 0 \times 0.01 + 3 \times 0.001$

SRB
118

4 Write a division number story that would give an answer of $\frac{3}{5}$.

Sample answer: A baker is using 3 cups of flour to make 5 loaves of bread. How many cups will she use in each loaf?

SRB
163-164

5 A clerk stacked these boxes to create a store display. Each box is 1 cubic unit. Find the volume of the store display.

SRB
231-232,
234

$V = $ _____54_____ units³

6 Write a division problem for which your estimate might be:

$7,000 \div 100 = 70$

_____ Answers vary. _____
_____ ÷ _____ → _____

Solve your problem:

SRB
84,
109-110

① 5.NF.2 ② 5.NBT.3, 5.NBT.3a ③ 5.NBT.1, 5.NBT.3, 5.NBT.3a
④ 5.NF.3 ⑤ 5.MD.5, 5.MD.5c ⑥ 5.NBT.6

137

Graphing Data as Ordered Pairs

The data in the table show Lilith's and Noah's ages at 5 different times in their lives.

SRB
55-56,
275

Lilith's Age (years)	Noah's Age (years)
5	1
7	3
9	5
11	7
12	8

Ordered Pairs:

(5 , 1)

(7 , 3)

(9 , 5)

(11, 7)

(12, 8)

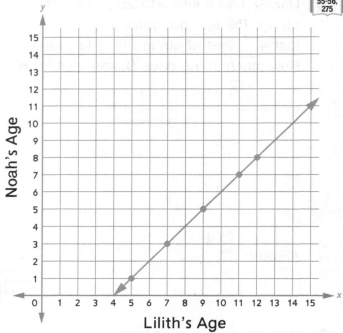

1. Write their ages as ordered pairs. Then plot the points on the grid.

2. What do you notice about the points you plotted?
   Sample answer: They are on a straight line.

3. Connect the points with a line. What other information can we get from this line?
   Sample answer: ages of Lilith and Noah at other times

4. Use the line to determine the ages of Lilith and Noah at various points in their lives.
   a. When Lilith was 8 years old, Noah was ____4____ years old.
   b. When Noah was 6 years old, Lilith was ____10____ years old.
   c. How old will Lilith be when Noah is 11? ____15____

5. Explain how you solved Problem 4c. Sample answer: I followed the y-axis up to 11 since Noah's age is the y-coordinate. Then I followed that grid line over until I hit the line I drew. It was 15 units over, so the x-coordinate would be 15. Lilith is 15.

6. Who is older, Lilith or Noah? ____Lilith____ How much older? ____4 years____

5.OA.3, 5.G.1, 5.G.2, SMP4, SMP7

Forming and Graphing Ordered Pairs

For each data set, fill in the missing values and write the data as ordered pairs. Plot the points on the grid and connect them using a straightedge. Use the graph to answer the questions.

SRB
55-56,
275

1 Dean is raising money for charity. He earns $2 for each lap he runs around the gym.

Laps Run (x)	$ Earned (y)
1	2
2	4
3	6
4	8

Ordered pairs:

(1 , 2)

(2 , 4)

(3 , 6)

(4 , 8)

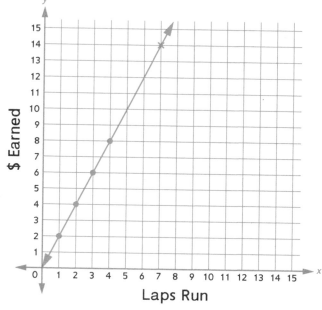

a. If Dean has earned $14, how many laps has he run? ____7____ laps

b. Put an X on the point on the grid that shows your answer to Part a.

c. What are the coordinates for this point? (__7__ , __14__)

2 Sally uses 2 paintbrushes for each paint jar.

Brushes (x)	Paint Jars (y)
2	1
4	2
6	3
8	4

Ordered pairs:

(2 , 1)

(4 , 2)

(6 , 3)

(8 , 4)

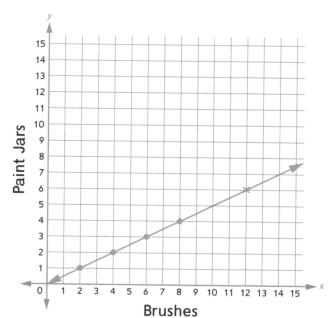

a. If Sally uses 6 jars of paint, how many brushes does she need? ____12____ brushes

b. Put an X on the point on the grid that shows your answer to Part a.

c. What are the coordinates for the point you marked with an X? (__12__ , __6__)

5.OA.3, 5.G.1, 5.G.2, SMP4, SMP7

139

Math Boxes

1 Write < or > to make true number sentences.

a. 0.5 _____<_____ 1.0

b. 3.2 _____>_____ 3.02

c. 4.83 _____>_____ 4.8

d. 6.25 _____<_____ 6.4

e. 0.7 _____>_____ 0.07

SRB 121-123

2 Write in standard notation.

$2 \times 10^3 =$ 2,000

$7 \times 10^5 =$ 700,000

$3 \times 10^2 =$ 300

SRB 68-69

3 Solve.

$\frac{1}{2}$ of 9 = $\frac{9}{2}$, or $4\frac{1}{2}$

$\frac{1}{4}$ of 5 = $\frac{5}{4}$, or $1\frac{1}{4}$

SRB 195

4 Write each decimal in words.

a. 0.16

_____Sixteen hundredths_____

_____

b. 3.28

_____Three and twenty-_____
_____eight hundredths_____

SRB 117

5 **Writing/Reasoning** Explain how you compared the decimals in Problem 1.

Sample answer: I looked at the first digit in the numbers and checked if they were in the same place-value positions. If the places were the same, then I looked at the value of the digits and compared those. I used that to figure out which number was greater.

SRB 121-123

① 5.NBT.3, 5.NBT.3b ② 5.NBT.2 ③ 5.NF.4, 5.NF.4a
④ 5.NBT.3, 5.NBT.3a ⑤ 5.NBT.3, 5.NBT.3b, SMP6, SMP7

140

Logo

Amy is designing a logo for her school club. She plans to put a trapezoid around the letters RC, which stand for Running Club. Below is the picture of the original trapezoid she drew on a coordinate grid.

SRB
55-56, 275

Original Trapezoid

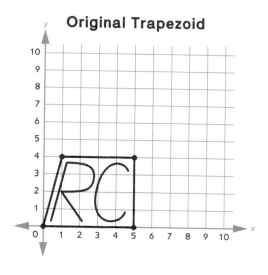

Amy decides she wants to include the school's name, so she needs to make the trapezoid wider. She does not want it to be taller. She developed a rule to help her fix the drawing.

Amy's Rule: Double the first coordinate of all the points.

1. If Amy uses her rule, what do you think the new trapezoid will look like? Why? Be specific in your description.

Sample answer: I think the trapezoid will be twice as wide, because I will multiply the x-coordinates by 2, and that will make the x-coordinates twice as large, which will move the points to the right. I do not think the height of the trapezoid will change, because Amy's rule does not change the y-coordinates.

5.G.1, 5.G.2, SMP2, SMP5

141

② Use Amy's rule to write the coordinates for the new trapezoid.

Original Trapezoid	New Trapezoid
(0, 0)	(0, 0)
(1, 4)	(2, 4)
(5, 4)	(10, 4)
(5, 0)	(10, 0)
(0, 0)	(0, 0)

③ Plot the new coordinates on the grid below. Connect the points in the same order you plot them.

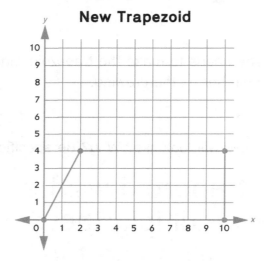

New Trapezoid

④ Does the new trapezoid look the way you expected? Why or why not? Be specific about how it changed.

Sample answer: The new trapezoid looks the way I expected. The new trapezoid is twice as wide as the original trapezoid. The x-coordinates determine how far to the right or left a point is, so now the x-coordinates that are not 0 are twice as far to the right because we multiplied them by 2. The height of the new trapezoid is the same as the original trapezoid because Amy's rule did not change the y-coordinates, which determine how far up or down the points are on the coordinate grid.

1 Use the unit squares to find the area of the rectangle.

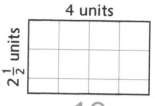

4 units

$2\frac{1}{2}$ units

Area = __10__ units²

SRB
224-225

2 Write 5 multiples of 7.

Sample answers:
7, 14, 21, 28, 35

SRB
72

3 Shoshana and Ariel got different answers on a fraction estimation problem. For the problem $\frac{8}{9} + \frac{1}{4}$:

Shoshana wrote $\frac{8}{9} + \frac{1}{4} > 1$.

Ariel wrote $\frac{8}{9} + \frac{1}{4} < 1$.

Who is correct? __Shoshana__

How do you know?

Sample answer: I know $\frac{8}{9}$ is $\frac{1}{9}$ less than 1, but $\frac{1}{4}$ is greater than $\frac{1}{9}$, so the sum of the two would be greater than 1.

SRB
181-182

4 Write all the factors of 30.

1, 2, 3, 5, 6, 10,
15, 30

SRB
73

5 Fill in the missing number.

a. $\frac{1}{2} = \frac{5}{\boxed{10}}$

b. $\frac{2}{3} = \frac{\boxed{8}}{12}$

c. $\frac{9}{\boxed{10}} = \frac{90}{100}$

SRB
166,
168-170

6 Solve.

a. If 4 is $\frac{1}{2}$ of the whole, what is the whole? __8__

b. If 2 is $\frac{1}{3}$ of the whole, what is the whole? __6__

SRB
195

① 5.NF.4, 5.NF.4b ② 5.NF.1 ③ 5.NF.2 ④ 5.NF.1 ⑤ 5.NF.1
⑥ 5.NF.4, 5.NF.4a

Math Boxes

Decimal Addition and Subtraction with Grids

For Problems 1 and 2: Sample shading is given.

SRB
129

- Shade the grid in one color to show the first addend.

- Shade more of the grid in a second color to show the second addend.

- Write the sum to complete the number sentence.

1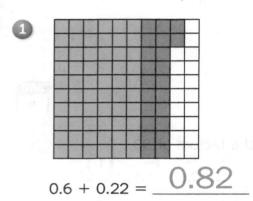

0.6 + 0.22 = **0.82**

2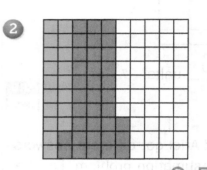

0.18 + 0.35 = **0.53**

For Problems 3 and 4: Sample shading is given.

- Shade the grid to show the starting number.

- Cross out or shade darker to show what is being taken away.

- Write the difference to complete the number sentence.

3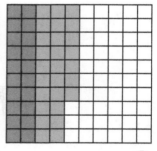

0.47 − 0.20 = **0.27**

4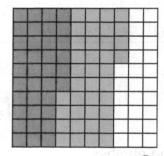

0.74 − 0.36 = **0.38**

5 Choose one of the problems above. Clearly explain how you solved it. Sample answer: In Problem 2, I shaded 1 column to show the 1 tenth in 0.18 and 3 columns to show the 3 tenths in 0.35. Then I shaded 8 squares for the 8 hundredths in 0.18 and 5 squares for the 5 hundredths in 0.35. There were 5 full columns and 3 more squares shaded, or 0.53 in all.

Math Boxes

1 Plot the following points on the grid.

    a. (1, 1) **b.** (2, 3)

    c. (5, 3) **d.** (4, 1)

    e. (1, 1)

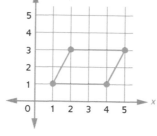

Connect the points in the order given.
What shape have you drawn?

Sample answer:
a quadrilateral

SRB
268, 275

2 Solve. You may use fraction circle pieces to help you.

    a. $\frac{1}{2} + \frac{1}{4} =$ $\frac{3}{4}$

    b. $\frac{1}{2} + \frac{3}{4} =$ $\frac{5}{4}$, or $1\frac{1}{4}$

    c. $\frac{3}{4} + \frac{1}{8} =$ $\frac{7}{8}$

    d. $\frac{1}{4} + \frac{3}{8} =$ $\frac{5}{8}$

SRB
166, 189

3 Make an estimate and then solve.

    Estimates vary.
        (estimate)

$$\begin{array}{r} 1\ 9\ 4 \\ *\ 2\ 1\ 5 \\ \hline 4\ 1{,}7\ 1\ 0 \end{array}$$

SRB
83,
100-104

4 Round to the nearest tenth.

    a. 45.52 = 45.5

    b. 60.18 = 60.2

    c. 123.45 = 123.5

    d. 38.27 = 38.3

    e. 56.199 = 56.2

SRB
124-127

5 **Writing/Reasoning** Explain why the order of the numbers in an ordered pair is important.

Sample answer: Order is important because the first number tells how far from 0 to move to the right, and the second number tells us how far from 0 to move up. If we don't follow the order, the point will be in the wrong place.

SRB
275

① 5.G.1 ② 5.NF.1 ③ 5.NBT.5 ④ 5.NBT.4
⑤ 5.G.1, SMP6, SMP7

Using Algorithms to Add Decimals

For Problems 1–6, make an estimate. Write a number sentence to show how you estimated. Then solve using partial-sums addition, column addition, or U.S. traditional addition. Show your work. Use your estimates to check that your answers make sense.

SRB 128, 130

---

① 2.3 + 7.6 = ?

Sample answer:
_____
(estimate)

$2 + 8 = 10$

2.3 + 7.6 = __9.9__

---

② 6.4 + 8.7 = ?

Sample answer:
_____
(estimate)

$6 + 9 = 15$

6.4 + 8.7 = __15.1__

---

③ 7.06 + 14.93 = ?

Sample answer:
_____
(estimate)

$7 + 15 = 22$

7.06 + 14.93 = __21.99__

---

④ 21.47 + 9.68 = ?

Sample answer:
_____
(estimate)

$21 + 10 = 31$

21.47 + 9.68 = __31.15__

---

⑤ 3.514 + 5.282 = ?

Sample answer:
_____
(estimate)

$3.5 + 5 = 8.5$

3.514 + 5.282 = __8.796__

---

⑥ 19.046 + 71.24 = ?

Sample answer:
_____
(estimate)

$20 + 70 = 90$

19.046 + 71.24 = __90.286__

---

⑦ Choose one problem. Answer the questions below.

a. How did you make your estimate? Sample answer: In Problem 5, I knew 3.514 was close to 3.5 and 5.282 was close to 5. I added those easy numbers to get my estimate: 3.5 + 5 = 8.5.

b. How did you use your estimate to check that your answer made sense?
Sample answer: I got 8.796 for my exact answer. That's only a little bit more than my estimate, 8.5, so my answer makes sense.

Math Boxes

① Put the following numbers in order from least to greatest.

7.1   7.01   0.0071   0.71

0.0071, 0.71, 7.01, 7.1
‾Least‾ ‾‾‾ ‾‾‾ ‾Greatest‾

SRB
121-123

② Write in exponential notation.

a. $30,000 = \underline{\quad 3 \quad} \times 10^{\boxed{4}}$

b. $6,000,000 = \underline{\quad 6 \quad} \times 10^{\boxed{6}}$

Sample answers given.

SRB
68-69

③ Evelyn is going on a hike. She will hike 4 miles in all. So far, she has hiked $\frac{1}{2}$ of the total distance. How far has she hiked?

Answer: ___2___ miles

SRB
195

④ Write the decimal in words.

a. 31.04

thirty-one and
four hundredths

b. 6.208

six and two hundred
eight thousandths

SRB
117

⑤ **Writing/Reasoning** Sarah said that 10^4 is the same as 40, because 10 * 4 is 40. Explain Sarah's mistake.

Sample answer: Sarah's mistake was that she multiplied 10 × 4, but she should have multiplied 10 × 10 × 10 × 10 and gotten 10,000.

SRB
68

① 5.NBT.3, 5.NBT.3b ② 5.NBT.2 ③ 5.NF.4, 5.NF.4a
④ 5.NBT.3, 5.NBT.3a ⑤ 5.NBT.2, SMP3

147

Using Algorithms to Subtract Decimals

For Problems 1–6, make an estimate. Write a number sentence to show how you estimated. Then solve using trade-first subtraction, counting-up subtraction, or U.S. traditional subtraction. Show your work. Use your estimates to check that your answers make sense.

SRB
128,
131-132

---

1 4.6 − 3.2 = ?

Sample answer:
_____
(estimate)

$5 - 3 = 2$

4.6 − 3.2 = __1.4__

---

2 13.1 − 8.7 = ?

Sample answer:
_____
(estimate)

$13 - 9 = 4$

13.1 − 8.7 = __4.4__

---

3 6.87 − 2.52 = ?

Sample answer:
_____
(estimate)

$7 - 2.5 = 4.5$

6.87 − 2.52 = __4.35__

---

4 24.07 − 12.68 = ?

Sample answer:
_____
(estimate)

$24 - 13 = 11$

24.07 − 12.68 = __11.39__

---

5 62.432 − 19.712 = ?

Sample answer:
_____
(estimate)

$62.5 - 20 = 42.5$

62.432 − 19.712 = __42.720__

---

6 17.41 − 6.274 = ?

Sample answer:
_____
(estimate)

$17 - 6 = 11$

17.41 − 6.274 = __11.136__

---

7 Choose one problem. Think about the algorithm you used. Answer the questions below.

a. How did your choice of algorithm help you get an accurate answer?

Sample answer: In Problem 1, I used trade-first subtraction. It helped me get an accurate answer because it helped me remember to check whether I needed to trade.

b. Was your choice of algorithm the most efficient choice? Why or why not?

Sample answer: I think trade-first subtraction was efficient for this problem because all I had to do was subtract each column. I didn't need to make trades, so there were no extra steps.

Math Boxes

1 Write the coordinates for each of the points on the coordinate grid.

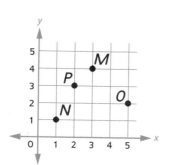

M: (__3__ , __4__)

N: (__1__ , __1__)

O: (__5__ , __2__)

P: (__2__ , __3__)

SRB
275

2 Solve. You may use fraction circle pieces to help you.

a. $\frac{1}{5} + \frac{3}{10} = \frac{5}{10}$, or $\frac{1}{2}$

b. $\frac{1}{5} + \frac{1}{10} = \frac{3}{10}$

c. $\frac{2}{3} + \frac{1}{6} = \frac{5}{6}$

SRB
166, 189

3 Write a multiplication problem for which your estimate might be:

$300 \times 70 = 21{,}000$

_____ × _____ = ?

Solve your problem.

Answers vary.

SRB
83,
100-104

4 Round to the nearest hundredth.

a. 67.467 = __67.47__

b. 9.017 = __9.02__

c. 43.284 = __43.28__

d. 16.107 = __16.11__

e. 5.658 = __5.66__

SRB
124-127

5 **Writing/Reasoning** Explain your strategy for rounding a decimal to the nearest hundredth in Problem 4.

Sample answer: Since the hundredths place is 2 places to the right of the decimal point, I know that each rounded number should have two places after the decimal point when rounded. I look at that digit and then use the digit right after to tell me if I should round the hundredths digit up to the next number or leave it as it is (round down).

SRB
124-127

① 5.G.1 ② 5.NF.1 ③ 5.NBT.5 ④ 5.NBT.4
⑤ 5.NBT.4, SMP6, SMP7

Math Boxes

Finding Areas of New Floors

Several rooms at Westview School will have new tile floors installed next year. Each tile is 1 square yard. For each room, find:

a. the number of tiles needed to cover the floor

b. the area of the floor in square yards

1 **Music Room**

$7\frac{1}{2}$ yd

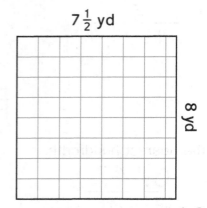

8 yd

a. Number of tiles: ___60___

b. Area: ___60___ square yards

2 **Office**

6 yd

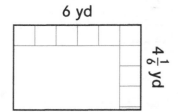

$4\frac{1}{6}$ yd

a. Number of tiles: ___25___

b. Area: ___25___ square yards

3 **Cafeteria**

12 yd

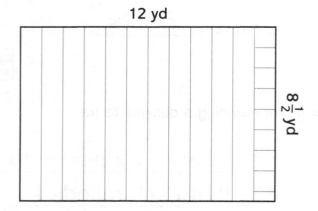

$8\frac{1}{2}$ yd

a. Number of tiles: ___102___

b. Area: ___102___ square yards

4 **Art Room**

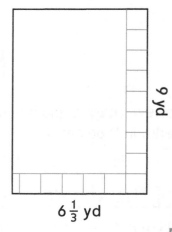

9 yd

$6\frac{1}{3}$ yd

a. Number of tiles: ___57___

b. Area: ___57___ square yards

5.NF.4, 5.NF.4b, SMP4

Math Boxes

Math Boxes

1 Write <, >, or =.

a. 0.90 __>__ 0.89

b. 3.52 __<__ 3.8

c. 6.91 __>__ 6.3

d. 4.05 __<__ 4.2

e. 0.38 __<__ 0.5

SRB
121-123

2 Which of the following shows *expanded* form using *exponential* notation for the number 315,796?

Fill in the circle next to the best answer.

Ⓐ 300,000 + 10,000 + 5,000 + 700 + 90 + 6

Ⓑ $3 \times 100{,}000 + 1 \times 10{,}000 + 5 \times 1{,}000 + 7 \times 100 + 9 \times 10 + 6 \times 1$

Ⓒ $3 \times 10^5 + 1 \times 10^4 + 5 \times 10^3 + 7 \times 10^2 + 9 \times 10^1 + 6 \times 10^0$

SRB
68-70

3 Solve.

a. $\frac{1}{3}$ of 30 = __10__

b. $\frac{1}{8}$ of 16 = __2__

c. $\frac{1}{5}$ of 25 = __5__

SRB
195

4 How would you write 54.279 in words? Choose the best answer.

 fifty-four and two hundred seventy-nine thousandths

◯ fifty-four and two hundred seventy-ninths

◯ fifty-four and two hundred seventy-nine hundredths

SRB
117

5 **Writing/Reasoning** Write a number story that could be modeled by Problem 3a.

Sample answer: There are 30 students in the school dance program. $\frac{1}{3}$ of them are in the advanced class. How many students are in the advanced class?

SRB
195

① 5.NBT.3, 5.NBT.3b ② 5.NBT.1, 5.NBT.2 ③ 5.NF.4, 5.NF.4a
④ 5.NBT.3, 5.NBT.3a ⑤ 5.NF.4, 5.NF.4a, SMP4

151

Math Boxes

1 How many unit squares cover the rectangle?

6 units

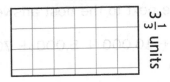

$3\frac{1}{3}$ units

__20__ unit squares

What is the area of the rectangle?

Area = __20 square units__

SRB
224-225

2 Write 4 multiples of 11.

__Sample answers:__
__88, 22, 55, 110, 1,100__

SRB
72

3 $\frac{3}{4} - \frac{1}{10}$ _____ $\frac{3}{8} + \frac{1}{12}$

Choose the best answer.

⬭ >

⬭ <

⬭ =

SRB
181-182

4 Write all the factors of 26.

__1, 2, 13, 26__

_____

SRB
73

5 Circle True or False.

a. $\frac{1}{3} = \frac{3}{12}$ True (False)

b. $\frac{7}{7} = \frac{11}{11}$ (True) False

c. $\frac{2}{5} = \frac{6}{15}$ (True) False

SRB
166,
168-170

6 Solve.

a. What is $\frac{1}{3}$ of 12? __4__

b. What is $\frac{1}{5}$ of 10? __2__

SRB
195

① 5.NF.4, 5.NF.4b ② 5.NF.1 ③ 5.NF.2 ④ 5.NF.1 ⑤ 5.NF.1

152   ⑥ 5.NF.4, 5.NF.4a